A CULTURAL HISTORY OF INSECTS

VOLUME 4

A Cultural History of Insects
General Editor: Gene Kritsky

Volume 1
A Cultural History of Insects in Antiquity
Edited by Gene Kritsky

Volume 2
A Cultural History of Insects in the Medieval Age
Edited by Gene Kritsky

Volume 3
A Cultural History of Insects in the Renaissance
Edited by Sherilyn G.F. Smith

Volume 4
A Cultural History of Insects in the Age of Enlightenment
Edited by Matthew S. Lehnert

Volume 5
A Cultural History of Insects in the Age of Industry
Edited by Carol M. Anelli and Susan W. Fisher

Volume 6
A Cultural History of Insects in the Modern Age
Edited by Robert K.D. Peterson

A CULTURAL HISTORY OF INSECTS

IN THE AGE OF ENLIGHTENMENT

VOLUME 4

Edited by Matthew S. Lehnert

BLOOMSBURY ACADEMIC
LONDON • NEW YORK • OXFORD • NEW DELHI • SYDNEY

BLOOMSBURY ACADEMIC
Bloomsbury Publishing Plc
50 Bedford Square, London, WC1B 3DP, UK
1385 Broadway, New York, NY 10018, USA
29 Earlsfort Terrace, Dublin 2, Ireland

BLOOMSBURY, BLOOMSBURY ACADEMIC and the Diana logo
are trademarks of Bloomsbury Publishing Plc

First published in Great Britain 2024

Series design by Raven Design
Cover image: Illustration of a flea from
Robert Hooke's *Micrographia* (1665) © Wikimedia

A catalogue record for this book is available from the British Library.

A catalog record for this book is available from the Library of Congress.

LCCN 2023029669

ISBN: HB: 978-1-3500-0308-8
Pack: 978-1-3500-0321-7
ePDF: 978-1-3502-6664-3
eBook: 978-1-3502-6665-0

Series: The Cultural Histories Series

Typeset by Integra Software Services Pvt. Ltd.
Printed and bound in Great Britain

CONTENTS

ILLUSTRATIONS

CONTRIBUTORS

Jennifer Angus, University of Wisconsin, Madison, Wisconsin, USA

Ellen M. Camerato, Michigan State University, East Lansing, Michigan, USA

Frank N. Egerton, Department of History, University of Wisconsin – Parkside, Kenosha, Wisconsin, USA

Robert Hamilton IV, Kent State University at Stark, North Canton, Ohio, USA

James N. Hogue, California State University, Northridge, California, USA

Daytona D. Johnson, Kent State University at Stark, North Canton, Ohio, USA

Gene Kritsky, Dean of Behavioral and Natural Sciences, Mount St. Joseph University, Cincinnati, Ohio, USA

Matthew S. Lehnert, Kent State University at Stark, North Canton, Ohio, USA

Ann M. Martinez, Kent State University at Stark, North Canton, Ohio, USA

Will K. Reeves, Western Compliance Assurance Branch USDA – APHIS – BRS –WCAB, Ft. Collins, Colorado, USA

Kristen E. Reiter, Stark State University, North Canton, Ohio, USA

SERIES PREFACE

This six-volume set, *The Cultural History of Insects*, is a unique examination of how insects have infested the human experience. The set is divided into six time periods: Antiquity, the Medieval Age, the Renaissance, the Age of Enlightenment, Age of Industry, and the Modern Age. Within each time period, eight topics are addressed: the state of entomological knowledge during the period; insects and disease; insects and food; insect products; insects' role in mythology and religion; insects as symbols; the portrayal of insects in literature and language; and the representation of insects in art. This structure is meant to allow readers to tailor their reading to their specific interests: for example, if insects' influence in the history of art is of primary interest, the reader can select the art chapter from each volume; if a reader wants to understand the impact of insects on a particular period in human history, then reading the volume on that time period would be appropriate.

The coverage of the topics presented is not encyclopedic. Presented are histories and case studies describing how insects influenced human culture and survival. The geographic focus is generally the Western Hemisphere, but details are included of geographic intersections in which cross-cultural influence plays an important role in various topics and time periods.

Insects were here before humans. They have influenced our evolution, eaten our food, provided us with raw materials, changed our history, and been the muses for our creativity. These six volumes illustrate how insects have helped to shape our culture over the past five thousand years.

Gene Kritsky,
Mount St. Joseph University,
Cincinnati, OH

Introduction

MATTHEW S. LEHNERT

The Age of Enlightenment (1600–1820) propagated modern science and issued the rise of contemporary entomology. This period is characterized by a cultural change in how science was viewed and how science was practiced, due in part to the scientific pioneers and philosophers of this era and the preceding age. How the Age of Enlightenment influenced scientific growth perhaps is best described by borrowing terminology from the field of evolutionary biology—the Age of Enlightenment is a *key innovation*—a trait (or, in this case, a cultural period) that facilitated the subsequent branching and radiation of species (here, giving rise to numerous branches of science and its subdisciplines that we see today). The Age of Enlightenment is summarized in the timeless book *Jurassic Park*, where it is stated:

> For four hundred years since Galileo [i.e., since the Age of Enlightenment], science has always proceeded as a free and open inquiry into the workings of nature. Scientists have always ignored national boundaries, holding themselves above the transitory concerns of politics and even wars. Scientists have always rebelled against secrecy in research, and have even frowned on the idea of patenting their discoveries, seeing themselves as working to the benefit of all mankind.
>
> (Crichton 1990)

It was the cultural atmosphere of the Age of Enlightenment that provided the fertile soil needed to nurture new ideas and allow for scientific growth, ultimately resulting in the most indispensable discovery in the following age, evolution by natural selection, by the eminent Charles R. Darwin (1809–1882).

This book explores the world of entomology during the Age of Enlightenment, a critical time period for the physical, astronomical, chemical, and biological sciences; however, what is it about the Age of Enlightenment that separates it from the preceding and following ages? It is probably impossible to provide an all-encompassing definition, as opinions about the period differ, but Immanuel Kant (1724–1804), a philosopher who lived during the Age of Enlightenment, stated "Enlightenment is man's emergence from nonage." He continued that the nonage is not due to a "lack on intelligence, but lack of determination and courage to use that intelligence without another's guidance. *Sapere aude*! Dare to know! Have the courage to use your own intelligence!" He also wrote "Are we now living in an enlightened Age? No, but we live in an age of enlightenment" and continues "what is more, the hindrances against general enlightenment or the emergence from self-imposed nonage are gradually diminishing" (Shachtman 2014). In other words, the Enlightenment represents a period of change, where new ideas and rationalistic thinking were overcoming the overbearing shackles of old traditions. Given Kant's statement, it is interesting to note that the people who lived during this age were well aware that the new culture was something special and distinct from the previous ages.

What is it though about the Age of Enlightenment that nourished the growth of science and the birth of entomology? The Introduction here is meant to provide a general overview of the political environment, the cultural atmosphere, its philosophers and scientists, and key events that nourished the growth of science, and entomology, during the Enlightenment, including the scientific revolution, which began near the end of the preceding age. These topics are far too expansive to be thoroughly covered in a single chapter, let alone in an entire book; however, a sample of the important events are discussed here.

A SCIENTIFIC REVOLUTION: ASTRONOMICAL REVELATIONS, UNIVERSAL PHYSICS, AND A NEW METHODOLOGY FOR PRACTICING SCIENCE

In order to better understand scientific growth and entomology during the Age of Enlightenment it is necessary to briefly encroach on the previous age and discuss the beginnings of the scientific revolution. Perhaps the most widely considered starting point for the scientific revolution is the publication *De revolutionibus orbium coelestium* (On the Revolutions of the Celestial Spheres) in 1543 by the astronomer Nicolaus Copernicus (1473–1543) (Repcheck 2008). Copernicus's removal of the Earth as the center of the universe—a concept originally proposed by Aristotle (384–322 BCE) in his book *On the Heaven*, popularized by Ptolemy (*c.* 150 CE) in *Mathematical Syntaxis*, and central dogma to the Catholic Church—not only employed quantitative measurements and observations, but also started a scientific crusade.

The concept that humans are not at the center of the universe, and that we are perhaps a rather narcissistic species to think so, was later expressed by the twentieth-century philosopher, Ludwig Wittgenstein, who asked a friend "Tell me, why do people always say it was natural for a man to assume that the sun went round the Earth rather than that the Earth was rotating?" His friend replied "Well, obviously because it just looks as though the Sun is going round the Earth." Wittgenstein replied "Well, what would it have looked like if it had looked as though the Earth was rotating?" (Anscombe 1959). This quote humorously exposes that humans are naturally egotistical and that thinking of us as the center of all things comes quite naturally. The quote also can be interpreted as pointing out that people should question widespread assumptions and imagine other scenarios that could provide equal explanation, which captures many of the opinions of philosophers and scientists during the Enlightenment.

Galileo Galilei (1564–1642), Johannes Kepler (1571–1630), and Francis Bacon (1561–1626) also made vital contributions to the scientific revolution early in the Age of Enlightenment. Galileo Galilei followed in Copernicus's footsteps and challenged religion with the heliocentric theory. His publication, *Sidereus Nuncius* (*Starry Messenger*) in 1610, which reported new astronomical observations, such as the phases of Venus and its orbit around the Sun, supported Copernicus's findings, thus further enabling the removal of Earth as the center of the universe. The Roman Inquisition concluded Galileo's support of heliocentrism as "foolish and absurd in philosophy, and formally heretical since it explicitly contradicts in many places the sense of Holy Scripture" (Sharratt 1994; Hannam 2009); therefore, instigating a critical movement for a separation between science and religion. Despite the Catholic Church's official statement, Galileo went on to continue promoting heliocentrism in his 1632 publication *Dialogue Concerning the Two Chief World Systems* and he consequently was put on house arrest in 1633 for the remainder of his life. Galileo is reported to have muttered "Eppur si muove" (nevertheless, it moves) as he left the courtroom, referring to the movement of celestial bodies (Livio 2020). The Vatican did finally overturn their verdict of Galileo in 1992, a mere 359 years after their initial sentence.

Galileo's work also employs novel methodology; he used the scientific method to acquire his results, therefore, he was able to promote the repeatability of his findings. The practice of the scientific method and the importance of repeatability is now central to entomological studies. However, even though Jesuit astronomers made the same observations of those of Galileo, the Church still refuted his claims and chose to remain blind to the evidence. Galileo was obviously frustrated by this, as expressed in his letter to Kepler in 1610, which stated:

> My dear Kepler, I wish that we might laugh at the remarkable stupidity of the common herd. What do you have to say about the principal philosophers of this academy who are filled with the stubbornness of an asp and do not want to look at either the planets, the moon or the telescope, even though I have freely and deliberately offered them the opportunity a thousand times? Truly, just as the asp stops its ears, so do these philosophers shut their eyes to the light of truth.
>
> (cited in Favaro 1900)

The German astronomer Johannes Kepler (1571–1630) made extraordinary advances about the physics of our solar system at the beginning of the Age of Enlightenment. Kepler, like Galileo, defended the workings of Copernicus, and was one of the first to do so in his 1596 declaration *Mysterium Cosmographicum*. Although his prescribed religion was Lutheran, Kepler did not strictly adhere to the Lutheran teachings, and refused to convert to Catholicism. His lack of a dedicated religion resulted in him having to frequently move during the Thirty Years War at the beginning of the seventeenth century. Despite him having to often relocate, Kepler went on to produce important works, including *Astronomia Nova* in 1609, *Harmonices Mundi* in 1619, and *Epitome Astronomiae* in 1621, where he outlined his laws of planetary motion. According to Gingerich (1992), Kepler's publication *Astronomia Nova* was truly significant because "it is the first published account wherein a scientist documents how he has coped with the multitude of imperfect data to forge a theory of surpassing accuracy."

Many of the fundamental scientific values that were propagated during the Enlightenment are still taught in modern entomology and introductory science courses. The philosopher Francis Bacon (1561–1626), for example, forever changed science in his book *Novum Organum, sive indicia vera de Interpretatione Naturae* (New Organon, or True Directions Concerning the Interpretation of Nature) in 1620 by providing the initial framework for how to properly conduct science, laying a foundation for what would become the scientific method. Bacon writes:

> There remains simple experience; which, if taken as it comes, is called accident, if sought for, experiment. The true method of experience first lights the candle [hypothesis], and then by means of the candle shows the way [arranges and delimits the experiment]; commencing as it does with

> experience duly ordered and digested, not bungling or erratic, and from it deducing axioms [theories], and from established axioms again new experiments.
>
> (Bacon 1902)

Bacon used empiricism as his groundplan for the scientific method, which is the idea that all knowledge comes from sense-experience; in other words, one cannot have knowledge about a subject if they did not experience it firsthand. The empiricism movement is well portrayed by the English philosopher and polymath John Locke (1632–1704), who stated "Whatever I write, as soon as I discover it not to be true, my hand shall be the forwardest to throw it into the fire" (Baldwin 1913).

Bacon's scientific method consisted of four steps in how to conduct science: to observe, measure, explain, and verify through replication. Although the scientific method has been improved upon since its origination (Castillo 2013), particularly with a greater emphasis based on the development of testable and falsifiable hypotheses, its initial foundation was based on inductive reasoning. As stated by Bacon, "If a man will begin with certainties, he shall end in doubts; but if he will be content to begin with doubts, he shall end in certainties" (Bacon 1605). To begin scientific inquiry with certainties was based on the Aristotelian theory, but Bacon felt another form of methodology needed to be developed. Bacon stated:

> [A]nother form of induction must be devised than has hitherto been employed, and it must be used for proving and discovering not first principles (as they are called) only, but also the lesser axioms, and the middle, and indeed all. For the induction which proceeds by simple enumeration is childish.
>
> (Bacon 1902)

The work by Hans Christian Ørsted (1777–1851) discussed the importance of inductive reasoning and stated:

> In order to achieve completeness in our knowledge of nature, we must start from two extremes, from experience and from the intellect itself. ... The former method must conclude with natural laws, which it has abstracted from experience, while the latter must begin with principles, and gradually, as it develops more and more, it becomes ever more detailed. Of course, I speak here about the method as manifested in the process of the human intellect itself, not as found in textbooks, where the laws of nature which

> have been abstracted from the consequent experiences are placed first because they are required to explain the experiences. When the empiricist in his regression towards general laws of nature meets the metaphysician in his progression, science will reach its perfection.
>
> (Ørsted 1799)

Bacon's set of principles for the scientific method (observation, measurements, explanation, and verification through replication) would eventually be applied to medicine and reveal some physicians' practices as quackery. The physician Elisha Perkins (1741–1799), for example, is best known for his invention of the Perkins Patent Tractor; a medical device apparently capable of resolving pain associated with inflammation, rheumatism, and painful areas of the face and head. The tractors, according to Perkins, were made of unusual metal alloys and could "draw off the noxious electrical fluid that lay at the root of suffering" (Miller 1935). Perkins would apply the three-inch metal rods to specific points of pain or pass the tractors over regions of pain, which would miraculously make the patient feel better (Delbourgo 2006). These expensive devices (which were actually made of steel and brass, not *unusual metal alloys*) would sell for five guineas, a steep price in 1796. Despite the cost, Perkins sold many tractors, claimed to have cured five thousand cases, and consumers of the product included prominent figures, such as George Washington.

In 1796, Abijah Richardson, a physician in Massachusetts, following Perkins's protocol for working the tractors, treated a woman prone to severe headaches with magnets. He had to resort to using the magnets, which were likely of similar shape to the tractors, because he did not own his own set of tractors. The pain, according to Richardson, "gradually abated & left her" and he believed the tractors " … relieved pain by attracting & conveying heat from the pained part" (Richardson 1802).

Some medical institutions refuted the healing abilities of the tractors; however, the physician Christopher Caustic supported the tractors and published four cantos that were first written by Thomas Green Fessenden in 1803, as requested by Perkins himself. Canto III reads:

> Say to the public all you can say,
> Of magic spells, and necromancy;
> That Perkins and his crew are wizards,
> Conceal'd in sanctimonious vizards.
>
> Say to the public all you can say,
> Of wonder-working power of fancy:
> Tell what *imagination's* force is
> In crows and infants, dogs and horses.

What though they say, why to be sure,
If we by Fancy's aid can *cure*;
Then why not use Imagination,
A *cheap* and *simple* operation?

That learn'd physicians pine with hunger,
The while a spruce young patent-monger
Contrives to wheedle simple ninnies,
And *tractorise* away *our* guineas.

(Fessenden 1804)

After Perkins's death in 1799 from yellow fever, which he acquired while trying to cure patients with the same illness, John Haygarth, a physician in Bath who specialized in rheumatic fever, tested the tractors on five of his patients. Instead of using metal tractors, however, Haygarth used tractors made of wood, which were fashioned to look similar to the Perkins metal tractors. Of the five patients tested, four of them experienced pain relief, and the following day he tested the same patients again, but this time he used metal tractors, and had similar results (Booth 2005). Haygarth published *On the Imagination as a Cause & as a Cure of Disorders of the Body* in 1800, which revealed the placebo effect, and the importance of using the scientific method to uncover the factuality of data and results.

The scientific method continued to evolve throughout the Age of Enlightenment and philosophers began to rationalize that the application of inductive *and* deductive reasonings are both important for the scientific method. Karl Popper (1902–1994) would later refine the scientific method into its current form by advocating falsifiable hypotheses as the foundation of science—in other words, if it's not falsifiable, it's not science. Although the scientific method will likely continue to be refined (Wellington 1994), particularly considering the large breadth of scientific fields, the foundation of the scientific method and its principles, laid by Bacon, continue to be employed in entomology today.

The paramount works by Sir Isaac Newton mark the end of the scientific revolution. Isaac Newton (1642–1727) was born on a farm in Woolsthorpe, England, the same year as the death of Galileo. He later enrolled at the University of Cambridge's Trinity College in 1661 where he studied a classical curriculum. Due to the onset of the Great Plague in 1665, Newton returned home and experimented with light and color, alchemy, developed the branch of mathematics known as calculus (independent of Gottfried Wilhelm Leibniz), and eventually determined the principles of gravity. Newton said of this productive period "I was in the prime of my age for invention, and minded mathematics and philosophy more than at any time since" (Edwards 1984).

Newton was elected to the Royal Society in 1672 because of his work with optics and telescopes. It wasn't until 1687, however, that Newton published *Philosophiae Naturalis Principia Mathematica* (Mathematical Principles of Natural Philosophy), which announced the mathematical mechanism of gravity and how it describes the elliptical orbits of the planets. Only Darwin's 1859 publication *On the Origin of Species by Means of Natural Selection, or the Preservation of Favoured Races in the Struggle for Life* can arguably be placed on the same tier of scientific significance as *Philosophiae Naturalis Principia Mathematica*, which is to astronomers and physicists as what Darwin's book is to biologists and entomologists. The workings of Newton and his physics would later be applied to entomology, as well as many of fields of the natural sciences.

THE PERILS OF ESTABLISHING PRIORITY, PART I: SIR ISAAC NEWTON AND ROBERT HOOKE

Newton's refrainment from quickly publishing his scientific achievements was intimately bound to one of the greatest rivalries in science, the conflict between Newton and Robert Hooke (1635–1703). Hooke's list of accomplishments (including the illustration of the flea from his book *Micrographia: or Some Physiological Descriptions of Minute Bodies Made by Magnifying Glasses with Observations and Inquiries Thereupon* that is shown on the cover of this book) challenges those of Newton, and some of these accomplishments are discussed throughout this book (see Chapters 1, 3, 6, and 8), but the rivalry with Newton is presented here.

In 1666, more than twenty years before the publication of *Philosophiae Naturalis Principia Mathematica*, Hooke gave lectures about the possible forces that act on the movement of celestial bodies to the Royal Society, where he was currently employed as the official experimenter, busily performing experiments based on his own research interests, but perhaps more often performing experiments for others of higher stature in the Royal Society (Nauenberg 1994). He stated in one of his lectures:

> I have often wondered, why the planets should move about the sun according to Copernicus's supposition, being not included in any solid orbs (which the ancients possibly for this reason might embrace) nor tried to it, as their center, by any visible strings; and neither depart from it beyond such a degree, nor yet move in a straight line, as all bodies, that have one single impulse, ought to do ...
>
> (Birch 1756)

Hooke continued to lecture about the movement of celestial bodies and stated:

> ... At which time also I shall explain a System of the World differing in many particulars from any yet known, answering in all things to the common Rules of Mechanical Motion: This depends on three Suppositions. First, That all Celestial Bodies whatsoever, have an attraction or gravitating power towards their own Center, whereby they attract not only their own part, and keep them from flying from them, as we may observe the Earth to do, but that they do also attract all the other Celestial Bodies that are within the sphere of their activity; and consequently that not only the Sun and Moon have an influence upon the body and motion of the Earth, and the Earth upon them, but that Mercury also Venus, Mars, Saturn and Jupiter by their attractive powers have a considerable influence upon its motion as in the same manner the corresponding attractive power of the Earth hath a considerable influence upon every one of their motions also. The second supposition is this, That all bodies whatsoever that are put into a direct and simple motion, will so continue to move forward in a straight line, till they are by some other effectual power deflected and bent into a Motion, describing a Circle, Ellipsis, or some other more compounded Curve Line. The third supposition is, That these attractive powers are so much the more powerful in operating by how much the neare the body wrought upon is to their own Centers. Now what these several degrees are I have not yet experimentally verified ...
>
> (Hooke 1674)

Hooke finished by stating:

> ... This I only hint at present to such as have ability and opportunity of prosecuting this Inquiry, and are not wanting of industry for observing and calculating, wishing heartily such may be found, having myself many other things in hand which I would first compleat and therefore cannot so well attend it. But this I durst promise the Undertaker, that he will find all the great Motions of the World to be influenced by this Principle, and that the true understanding thereof will be the true perfection of Astronomy ...
>
> (Hooke 1674)

It was during this time that Isaac Newton was actively pursuing his interests in the physics of light. His observations of the behavior of light when it passes through prisms and the colors produced led him to conclude that light is composed of particles. To support this idea, he developed the first telescope that used reflective mirrors, replacing the lenses that cause light to disperse, which resulted in different colors focusing on different planes (i.e., chromatic aberration). Newton was asked to demonstrate his telescope to the Royal Society in 1671, which was followed by him publishing his notes, *Of Colours*. This body of work, however, was relentlessly criticized by Hooke, and unfortunately, Newton could not handle the criticism and retreated from public scientific debates (Worthington 2016).

Nearly a decade later, Hooke and Newton began to exchange letters when Hooke became the secretary of the Royal Society in 1679 and was tasked to correspond with other scientists (Turnbull 1960). His supposed purpose was to determine the research efforts of the other members of the Royal Society, including Isaac Newton; however, some scholars have suggested that Hooke was indirectly asking Newton for assistance with the complex mathematics that would be required to understand the physics of gravity (Nauenberg 1994). In his letters to Newton, Hooke asked Newton about his thoughts and interests on such matters as "... compounding the celestiall motions of the planetts of a direct motion by the tangent and an attractive motion towards the centrall body ..." Hooke, already having established a well-known interest in the movements of celestial bodies, had not yet developed the mathematics or experimental evidence to explain the mechanism of gravity.

Isaac Newton responded to Hooke's letter with a schematic of an orbit using the assumption of a constant radial force, which differed from Hooke's mathematical assumptions. From this response, Hooke assumed that Newton had indeed developed a mathematical principle for the motion of planets, and wrote a new letter to Newton stating:

> ... But my supposition is that the Attraction always is in a duplicate proportion to the Distance from the Center Reciprocall, and Consequently that the Velocity will be in a subduplicate proportion to the Attraction and Consequently as Kepler Supposes Reciprocall to the Distance ....
>
> (qtd. in Turnbull 1960)

In other words, Hooke supposed, as did others, including the astronomer Edmond Halley (1656–1742), that gravity might follow an inverse square law; however, he importantly misquoted Kepler, which would later be used by Newton as evidence that Hooke did not understand gravity. Hooke continued in his letter that "... I doubt not but that by your excellent method you will easily find what that Curve must be, and its proprietys, and suggest a physicall Reason of this proportion ..." Newton did not immediately respond to this letter (Turnbull 1960).

Hooke claimed to Edmond Halley in 1684 that he could provide the mathematical proof that an inverse square law would result in an elliptical orbit; however, that challenge never came to fruition. As stated by DeMoivre in 1727:

> ... in 1684 Dr. Halley made Sir Isaac a visit at Cambridge [after he had visited Hooke] and there in a conversation the Dr. asked him what he thought the Curve would be that would be described by the Planets supposing the force of attraction towards the Sun to be reciprocal to the square of their distances from it. Sir Isaac replied immediately that it would be an *Ellipsis*. The Docter struck with joy and amazement asked him how he knew it. Why saith he I have calculated it. Whereupon Dr. Halley asked him for his calculation without any further delay. Sir Isaac looked among his papers but could not find it, but he promised him to renew it, and then to send it to him ...
>
> (qtd. in Turnbull 1960)

Newton provided Edmond Halley a copy of the famous manuscript, *De Motu*, a couple of months later that contained the mathematical proof, which also included some illustrations and geometrical mechanisms that were similar to what Robert Hooke had previously sent in a letter to Newton. Newton, of course, would go on to publish *Philosophiae Naturalis Principia Mathematica* two years later, where he stated his three laws of motion: 1) every object in a state of uniform motion will remain in that state of motion unless an external force acts on it; 2) force equals mass times acceleration; and 3) for every action there is an equal and opposite reaction. Some of these laws, however, are very similar to the suppositions that Hooke had outlined in his 1666 lectures.

Isaac Newton did not acknowledge that Hooke provided any contribution to his discovery. Hooke apparently was enraged and felt he deserved co-authorship on Newton's work on gravity, but Newton provided evidence that Hooke's assumption of the inverse square law was already widespread and therefore did not give Hooke any credit. Also, Newton would use the letter that Hooke wrote to him with the misquoted statement as evidence that Hooke clearly did not understand the physical principals of gravity. Newton stated of Hooke:

> Yet am I not beholden to him for any light into that business but only for the diversion he gave me from my other studies to think on these things & for his dogmaticalness in writing as if he had found the motion in the Ellipsis, which inclined me to try it.
>
> (Turnbull 1960)

Newton continues to tell Haley in the letter (dated June 20, 1686):

> ... Borell did something in it and wrote modestly, he [Hooke] has done nothing and yet written in such a way as he knew and had sufficiently hinted all but what remained to be determined by the drudgery of calculations and observations, excusing himself from that labour by reason of his other business: whereas he should rather have excused himself by reason of his inability. For it is plain by his words he knew not how to go about it. Now is not this very fine? Mathematicians that find out, settle and do all the business must content themselves with being nothing but dry calculators and drudges and another that does nothing but pretend and grasp at all things must carry away all the invention as well of those that were to follow him as of those that went before ... For as Borell wrote long before him that by a tendency of the Planets towards the sun like that of gravity or magnetism the Planets would move in Ellipse, so Bullialdus wrote that all force respecting the sun as its center and depending on matter must be reciprocally in duplicate ratio of the distance from the center ...
>
> (Turnbull 1960)

The feud between Hooke and Newton would continue until Hooke's death in 1703, when Newton would take over as president of the Royal Society. Newton used Hooke's death as an opportunity to finally publish his studies on the physics of light in the book *Opticks* in 1704, because the unabated criticism by Hooke would no longer be a problem. Newton also is said to be responsible for the destruction of the only known portrait of Hooke, either through ordering to have it destroyed or *forgetting* to move it when the Royal Society relocated; however, there is a lack of evidence for either of these events. Newton would later be knighted and would go on to become considered one of the greatest scientists of all time. It was famously stated by the English poet Alexander Pope "Nature and nature's laws lay hid in night; God said 'Let Newton be' and all was light" (Sharp 2001).

As suggested by the French astronomer and, later guillotined, political leader Jean Sylvain Bailly (1736–1793), the scientific revolution consisted of two parts that included moving on from the old and launching the new (Cohen 1976). The first part of the scientific revolution was the recovery of the teachings and the knowledge of the ancients, which had been lost or abandoned for centuries prior to resurfacing during the Renaissance, and the work of Copernicus and Galileo (Cohen 1965). The second part and the completion of the scientific revolution, the establishment of the new, is represented by the world embracing the physics of universal gravity, and that a common mechanism describes the movements of the solar system, as outlined by the works of Isaac Newton.

THE AGE OF ENLIGHTENMENT: A PLETHORA OF POLYMATHS

The above forementioned pioneers and representatives of the scientific revolution share several common traits, for example, an extensive interest in the physics of the cosmos and of the natural sciences, but they also were polymaths, or people that are extremely knowledgeable—usually considered masters—in a variety of disciplines. The term polymath is applied to several people throughout the ages, including Aristotle, Leon Battista Alberti (1404–1472), and Leonardo da Vinci (1452–1519), who was described by Johnston and Smith (2003) as "prodigious polymath ... Painter, sculptor, engineer, astronomer, anatomist, biologist, geologist, physicist, architect, philosopher, humanist." The Age of Enlightenment, however, is unique in that the term polymath can refer to an extraordinarily large number of individuals.

The Enlightenment is often referred to as the Age of Reason and is characterized by rationalism. A period branded by elite philosophers, also considered polymaths, including Voltaire, Jean-Jacques Rousseau, and John Locke, the Age of Enlightenment boomed with a climate where observations and rationalism were celebrated, rather than simply accepting tradition or the unobserved as truth. Indeed, the great philosophers act as a symbol for this Age. As stated by Peter Gay:

> [The philosophers] were united on a vastly ambitious program, a program of secularism, humanity, cosmopolitanism, and freedom, above all, freedom in its many forms—freedom from arbitrary power, freedom of speech, freedom of trade, freedom to realize one's talents, freedom of aesthetic response, freedom, in a word, of a moral man to make his own world.
>
> (Gay 1969)

There is no doubt that the philosophers and their writings had an enormous impact on science, but science equally impacted the philosophers, thus resulting in a partnership where many of the great philosophers were scientists. The marriage between science and the philosophers was based on common goals—to use rationalism and methodology to unveil truth, and its perhaps the system of duality that promoted the rise of so many polymaths. The French philosopher Francois-Marie Arouet (1694–1778) (better known by his *nom de plume*, Voltaire), for example, was heavily influenced by Isaac Newton (and the first to publish the famous story of the apple falling on Newton's head [true or not], which he learned from Newton's niece (Pearson 2005), even performing some of his own experiments with optics.

One common theme among the great scientists, philosophers, and polymaths during the Enlightenment is that many were considered criminals because their writings often conflicted with the government and the church, sometimes overturning the former and often weakening the latter. Voltaire, for example, openly criticized French rule and Christianity, writing in a letter in 1767 to the King of Prussia, Frederick II:

> Ours [i.e., the Christian religion] is assuredly the most ridiculous, the most absurd and the most bloody religion which has ever infected this world. Your Majesty will do the human race an eternal service by extirpating this infamous superstition, I do not say among the rabble, who are not worthy of being enlightened and who are apt for every yoke; I say among honest people, among men who think, among those who wish to think. ... My one regret in dying is that I cannot aid you in this noble enterprise, the finest and most respectable which the human mind can point out.
>
> (Mathews 2009; Coakley 2012)

Voltaire's lack of ability to refrain from speaking his mind and writing about his religious and political views resulted in him being imprisoned in the Bastille (in a windowless cell with ten-foot-thick walls) and having to leave Paris twice. He first had to leave Paris in 1726 due to a conflict between him and the French Nobleman, the chevalier de Rohan-Chabot, who insulted Voltaire because of his name change. Voltaire's retorts resulted in de Rohan having him assaulted and later imprisoned without trial. Rather than him having to stay in prison for an unknown period, Voltaire suggested he leave France for England, which was approved (Davidson 2012). He was greatly influenced by the cultural differences between the two countries, particularly in government and religion. Voltaire's second exile from Paris was caused by his publication *Letters Concerning the English Nation* in 1733. This book, which was published without the approval of the royal censor, revealed to its readers in France how the government in England was perhaps superior in many respects, including human rights. This publication was banned and was followed by public burnings, and resulted in Voltaire moving to Cirey-sur-Blaise, where he had a long-time affair with the celebrated mathematician Emilie du Chatelet (1706–1749) (Mitford 2012).

Voltaire and du Chatelet both were greatly inspired by Isaac Newton, actively pursuing their own experiments and progressing Newton's work, and brought Newton's science to the public. In 1735, du Chatelet went on to publish the authoritative translation of Newton's *Principia* in French, and Voltaire's book *Elements of Newton's Philosophy* made Newton's complex mathematics and principles accessible to the public.

Voltaire, like many of the great philosophers and polymaths of the time, was a deist (Betts 1984), which means he believed in a higher power or supreme being that created the universe but does not intervene in the daily lives of people. It is a method of combining the belief of a higher power with rationalism. Voltaire stated "What is faith? Is it to believe that which is evident? No. It is perfectly evident to my mind that there exists a necessary, eternal, supreme, and intelligent being. This is no matter of faith, but of reason" (Murtagh and Lee 2017). He also promoted for people of different faiths to tolerate one another, but he certainly had a disdain for Christianity and particularly priests, stating that they "rise from an incestuous bed, manufacture a hundred versions of God, then eat and drink God, then piss and shit God." Many of Voltaire's views on religions were published as a collection of articles in *Dictionnaire philosophique* in 1764, one of his most famous pieces of works.

Voltaire, similar to his contemporaries, was outspoken about his opinions on multiple cultural matters. For example, he stated "at what price we eat sugar in Europe," referring to the atrocities of slavery, and "common sense is not so common," which is just as relevant now as it was then. Many of Voltaire's works and publications, which include more than twenty thousand letters, numerous and successful plays, novels, including science fiction works, and pamphlets not only influenced those that were alive during this time, but still influence writers today. Victor Hugo (1802–1185), writer of the novel *Les Miserables* in 1862 and *The Hunchback of Notre-Dame* in 1831, stated "To name Voltaire is to characterize the entire eighteenth century."

Few philosophers had more impact on government than the Genevan Jean-Jacques Rousseau (1712–1778), who was deeply influenced by Voltaire, and stated:

> The first man who, having fenced in a piece of land, said "This is mine", and found people naïve enough to believe him, that man was the true founder of civil society. From how many crimes, wars, and murders, from how many horrors and misfortunes might not any one have saved mankind, by pulling up the stakes, or filling up the ditch, and crying to his fellows: Beware of listening to this impostor; you are undone if you once forget that the fruits of the earth belong to us all, and the earth itself to nobody.
>
> (Rousseau 1755)

Rousseau's paramount political works include the *Discourse on Inequality* in 1754 and *The Social Contract* in 1762. Both books discuss the problems in society and how to establish a political community to address these problems. Although he was an advocate for Voltaire, often defending him, the gesture was not often reciprocated by Voltaire, who stated of Rousseau's *Discourse on Inequality*:

> No one has ever employed so much intellect to persuade men to be beasts. In reading your work one is seized with a desire to walk on four paws. However, as it is more than sixty years since I lost that habit, I feel, unfortunately, that it is impossible for me to resume it.
>
> (qtd. in Durant and Durant 1967)

Voltaire and Rousseau, although having some differences in opinion, both made enormous contributions to the French Revolution, which is considered one of the biggest political events in recent history.

The movements for deism, separation between church and state, and the general disdain for monarchy rule was not confined to enlightened European philosophers. The American Colonies also had great philosophers and polymaths, including Benjamin Franklin (1706–1790) and Thomas Jefferson (1743–1826), who were directly influenced by the views of the European philosophers. In fact, the culture that defined the Age of Enlightenment would directly impact and forge what would become the new government of the United States of America in 1776 (Staloff 2007). Benjamin Franklin, a polymath and freemason, even inspired Voltaire to join the freemasons and was present during his initiation in Paris in 1778 (Paterson 1956). Thomas Jefferson, like Voltaire and other European philosophers, was a leader of the movement of separation between church and state. Some Christians took his attitude as a sign of atheism and were against him serving as an elected official, which was expressed by the *New England Palladium* during his presidential campaign in 1800:

> Should the infidel Jefferson be elected to the Presidency, the seal of death is that moment set on our holy religion, our churches will be prostrated, and some infamous "prostitute", under the title of goddess of reason, will preside in the sanctuaries now devoted to the worship of the most high.
>
> (qtd. in Dreisbach 2003)

Of course, this was not a true statement, and Jefferson would go on to make a tremendous impact on the United States. The late president John F. Kennedy, hosting a group of Nobel laureates at the White House, famously stated "the most extraordinary collection of talent, of human knowledge, that has ever been gathered together at the White House—except when Thomas Jefferson dined alone" (Rees 2003).

Polymaths during the Age of Enlightenment also include Athanasius Kircher (1602–1680), who is said to be a "Master of a Hundred Arts" (Reilly 1974) and "The Last Man who Knew Everything" (Findlen 2004), Mikhail Vasilyevich Lomonosov (1711–1765), a Russian poet who also discovered the atmosphere of Venus and the Law of Mass Conservation, and published famous geological works, such as "On The Strata of the Earth" in 1763 (Chorley and Beckinsale 1991), Alexander von Humboldt (1769–1859), who is famous for his travels of the world, proposed that South America and Africa were once joined, and is the first person to describe human-induced climate change (Wulf 2015), and Blaise Pascal (1623–1662), who as a teenager invented the mechanical calculator, and later at the age of sixteen wrote the treatise on projective geometry (Editors 2018).

Another exceptional polymath during the Age of Enlightenment is Gottfried Wilhelm (von) Leibniz (1646–1716). Gottfried Leibniz was a German mathematician, philosopher, physicist, and "... perhaps the last great Renaissance man who in Bacon's words took all knowledge to be his provide" (Eliot 1871). A child prodigy who entered the University of Leipzig in 1661 at the age of fourteen and graduated with his first bachelor's degree in philosophy one year later, Leibniz would go on to earn a master's degree in 1664, another bachelor's degree in law in 1665, and then his Doctorate in Law in 1666. Later, Leibniz would meet the mathematician Christiaan Huygens, who took Leibniz as an apprentice to teach him physics and mathematics. It was through his self-studies that Leibniz created the first calculating machine capable of adding, subtracting, multiplying, and dividing, which resulted in him becoming an external member of the Royal Society in 1670. It was also during this time period that Leibniz would co-invent calculus, the field of mathematics most often attributed to Sir Isaac Newton, which led to the *other* arguably most famous historical dispute among scientists, again involving Sir Isaac Newton (Bardi 2006).

THE PERILS OF ESTABLISHING PRIORITY, PART II: SIR ISAAC NEWTON AND GOTTFRIED WILHELM (VON) LEIBNIZ

The notorious calculus controversy between Leibniz and Newton seems to be based rather on a series of misunderstandings fueled by others, such as Newtonians (i.e., followers of Newton), which then grew into a full-blown conflict. Leibniz was considered, at best, a novice mathematician before his tutelage under Christiaan Huygens began in 1672. By the time Leibniz visited London the following year and met Henry Oldenburg (Secretary of the Royal Society) and John Collins (mathematician, and also affiliated with the Royal Society), which furthered inspired his interests in series and infinitesimal mathematics, he had already made great strides in mathematics and would begin his work on developing calculus in 1674, which he would later publish in 1684 in his paper *Nova Methodus pro Maximus et Minimis* (Bardi 2006).

Isaac Newton began working on calculus in 1666, eight years before Leibniz began his own exploration of infinitesimal mathematics, but as mentioned before, Newton was profoundly impacted by his interactions with Robert Hooke and was no longer eager to publish. The evidence of when Newton developed a method for calculating fluxions (fluxion is Newton's term for "derivative") is found in his unpublished notes, and there seems to be no argument about this fact. However, Leibniz and most of the rest of the world, were completely unaware about Newton's tremendous efforts, and Leibniz was of the impression that Newton (who he had not yet met) was an expert in the method of series (Hall 2002). Newton would go on to publish *Philosophiae Naturalis Principia Mathematics* in 1687, where he provides his first account of calculus. Given this information it would seem that Leibniz deserves the credit for inventing calculus, because he was the first to publish it; however, the cultural atmosphere of science, particularly in determining correctly accrediting priori of discovery, was different then than it is now, and credit was often attributed to who had the "idea" first, not necessarily who was first to publish. In addition, there are several events that occurred before Leibniz's initial publication, which is what resulted in him being charged with plagiarism and wrongly discredited in the scientific community.

Collins and Oldenburg had a responsibility of communicating to the Royal Society members updates and discoveries among its scientists. In the spring of 1675, Collins and Oldenburg distributed a report about English mathematics to the Royal Society community, including Leibniz, which included some rather vague updates about Newton's current mathematical workings, but lacked any sufficient details about his work with fluxions. Given the information in this report, it is assumed that Leibniz was under the impression that Newton was considered an expert of mathematical series. Collins and Oldenburg, however, were aware of Newton's work on fluxions and also knew that Leibniz was now working with infinitesimal mathematics, and therefore encouraged Newton to publish first. Newton was not ready to publish and was persuaded by Oldenburg to at least start a correspondence with Leibniz, with Oldenburg acting as the mediator. This communication resulted in two famous letters from Newton to Leibniz, one written in June of 1676 now known as *Epistola prior*, and another in October of 1676 known as *Epistola posterior* (Turnbull 1960).

The first letter, the eleven-page *Epistola prior*, provided some details about Newton's work with binomial theorem and begins with:

> Though the modesty of Mr Leibniz ... pays great tribute to our countrymen for a certain theory of infinite series about which there begins to be some talk, yet I have no doubt that he has discovered not only a method for reducing any quantities whatever to such series, as he asserts, but also various shortened forms ... Since however he very much wants to know what has

> been discovered in this subject by the English, and since I myself fell upon this theory some years ago, I have sent you some of those things which occurred to me in order to satisfy his wishes ...
>
> (qtd. in Turnbull 1960)

As promised, Leibniz responded to Newton (through Oldenburg) and provided some insight into the mathematics of determining the infinite sum for $\pi/4$ and asked for some explanations regarding Newton's calculations of series. Newton, concerned that someone might scoop his work on fluxions, responded with the nineteen-page *Epistola posterior*, which was accompanied with a cover letter only for Oldenburg (Turnbull 1960). Newton stated in this cover letter:

> I can hardly tell with what pleasure I have read the letters of those very distinguished men Leibniz and Tschirnhaus. Leibniz's method for obtaining convergent series is certainly very elegant, and it would have sufficiently revealed the genius of its author, even if he had written nothing else. But what he has scattered elsewhere throughout his letter is most worthy of his reputation, it leads us to hope for very great things from him. The variety of ways by which the same goal is approached has given me the greater pleasure, because three methods of arriving at series of that kind had already become known to me, so [*scribbled-out words*].
>
> (qtd. in Turnbull 1960)

The [*scribbled out words*] shown here refers to the heavily scribbled-out writing that Newton wanted to conceal, which was later found in Newton's own notes to be admittance that he was previously unfamiliar with Leibniz's series for $\pi/4$. He also expressed to Oldenburg that:

> I hope this will so far satisfy M. Leibniz that it will not be necessary for me to write any more ... For having other things in my head, it proves an unwelcome interruption to me to be at this time put upon considering these things.
>
> (qtd. in Turnbull 1960)

Newton clearly did not want to reveal the scope of his previous work with fluxional calculus. In his letter to Leibniz he writes "I cannot proceed with the explanation of the fluxions now, I have preferred to conceal it thus: 6accdae13eff7i319n4o4qrr4s8t12vx" (Turnbull 1960). This is an anagram; which Newton would use later to further establish his priority over calculus. The anagram requires decoding and refers to how many of each letter are found in a statement about his principal theorem. Once decoded, it states in

Latin "Data aequatione quotcunque fluentes quantitates involvente, fluxiones invenire: et vice versa," which translated in English means "Given an equation involving any number of fluent quantities, to find the fluxions: and vice versa" (Schrader 1962).

Newton increasingly became critical of Oldenburg and Collins and their sharing of information about what Newton had been working on. Newton wrote to Oldenburg two days after his letter to Leibniz stating "Pray let none of my mathematical papers be printed without my special license" (Bardi 2006). Newton also wrote to Collins telling him to let his mathematical studies "ly by till I am out of the way." Unfortunately, by this point in time Collins had already provided Leibniz access to Newton's private papers in possession of Collins, apparently not thinking anything ill could have come of it. Naturally, Collins did not report back to Newton that Leibniz was provided some access to his papers, but this would be brought to light later.

Leibniz published his methods of calculus in the journal *Acta Eruditorum* in 1684, which is a monumental achievement that establishes many of the principles and symbols we still use in modern calculus; however, this publication did not provide any credit to, nor even mention, Isaac Newton. Leibniz would publish again two years later, also not mentioning his correspondence with Newton. *Philosophiae Naturalis Principia Mathematica* was published in 1687, but Newton only published a part of his work with calculus and did refer to his initial letters with Leibniz in a passage. Later editions of *Philosophiae Naturalis Principia Mathematica* mentioned Newton's letters to Collins in 1672; a move for Newton to claim priority (Schrader 1962).

Apparently, Newton and Leibniz, although not in correspondence during this period of activity, were not quarreling about the issue of priority of the calculus. Leibniz was content accepting that his calculus and that of Newton's was a finding through convergence, which is supported by the fact that their methods of calculus are very different from one another, yet both result in the same answer. Newton also didn't immediately seem perturbed by Leibniz's publications. However, the publication of *Philosophiae Naturalis Principia Mathematica* brought him fame and fortune, thus bringing his work on the calculus to the forefront of his fame because of its applications to many different fields.

The dispute between Newton and Leibniz began to take shape in 1693, when the mathematician John Wallis published his revised edition of *Algebra*. In this book, Wallis briefly presents Newton's work on fluxional calculus. Wallis followed up this work with another publication in 1695, *Mathematical Works*, where he states that he learned the fluxional calculus through the letters from Newton to Leibniz, letters that actually lacked any sufficient details about calculus, thus making a false claim that would go on to be interpreted as stating that Leibniz learned calculus through the 1676 letters from Newton. Wallis was considered someone of extreme English nationalism who thought little of Germans and was concerned that the rest of Europe would surpass the mathematicians of England. For example, he wrote to Newton:

> ... with great applause by the name of Leibniz's Calculus Differentialis ... You are not so kind to your Reputation (& that of the Nation) as you might be when you let things of worth ly by you so long, till others carry away the Reputation that is due to you.
>
> (qtd. in Hall 2002)

At this point in the story people of stature are beginning to take sides as to who is the rightful inventor of calculus. Johann Bernoulli (1667–1748), for example, found Leibniz's publications of calculus incredibly useful in his own studies and wrote to Leibniz that perhaps Newton had plagiarized Leibniz's work. In 1699, the mathematician Nicolas Fatio de Duillier would move forward with outwardly accusing Leibniz of plagiarism (Hall 2002), and stated:

> But I now recognize, based upon the factual evidence that Newton is the first inventor of this calculus, and the earliest by many years; whether Leibniz, the second inventor, may have borrowed anything from him, I should rather leave to the judgement of those who have seen the letters of Newton and his other manuscripts. Neither the silence of the more modest Newton, nor the remitting exertions of Leibniz to claim on every occasion the invention of the calculus for himself, will deceive anyone what examines these records as I have.
>
> (qtd. in Hall 2002)

Both Leibniz and Newton were not intimately involved with the growing allegations against them.

Things took a turn for the worse when Newton, now the president of the Royal Society, published a more complete account of calculus, *Tractatus de Quadratura Curvarum*, as an appendix in his 1704 book *Opticks*. Here, Newton put forth the full power and potential of calculus, something of which Leibniz was already aware of, and Newton's methods of calculus were very similar to Leibniz's. Leibniz became increasingly suspicious of Newton's claims of priority, something he was previously content with sharing with Newton, but now suspected Newton of stealing from him. An anonymous review of Newton's work, which is well established to be written by a distressed Leibniz, stated:

> Accordingly, instead of the Leibnizian differences, Mr Newton employs, and has always employd, fluxions, ... He has made elegant use of these both in his *Principia Mathematica* and in other publications since, just as Honoré Fabri [a famous plagiarist of the time] in his *Synopsis Geometrica* substituted the advance of movements for the method of Cavalieri.
>
> (qtd. in Sonar 2018)

In other words, the anonymous reviewer (most likely Leibniz) accused Newton of plagiarism.

The Newtonian, John Keill, would later publish his paper "On the Laws of Centripetal Force" in the *Philosophical Transactions of the Royal Society* in 1710, where he outright lays Newton's claim to calculus. Leibniz was naturally offended by this statement and wrote to the Royal Society requesting a statement of apology. The statement, written by Keill with Newton's assistance and approval, stated:

> ... as he [referring to Newton] has in two letters to Oldenburg (which the latter transmitted to Leibniz) given pretty plain indications to that man of the most perceptive intelligence, whence Leibniz derived the principles of that calculus or at least could have derived them ...
>
> (qtd. in Hall 2002)

The letter also stated that some of Newton's work of calculus was provided to Leibniz via their correspondences many years earlier, and that he [Keill] would not formally press any criminal charges of plagiarism.

An angry Leibniz responded to Newton in 1711 and claimed that "everyone had received his due" regarding the discovery of calculus and for Keill to back down. Considering the charges at hand for priority of calculus and that there were quarrels among the members of the Royal Society over priority, Newton decided to have a formal committee created in 1712 that would oversee the matter and determine who has the rightful claim. Interestingly, according to Newton this committee consisted of an impartial constituency; however, the members of the committee members were not revealed until 1846, 134 years later, and found to consist entirely of Newton supporters and Newton himself. Evidence against Leibniz included his previous access to Newton's unpublished papers that were in the possession of Collins. Although true, we now know today that Leibniz had already developed his calculus before having access to these papers (Hall 2002). Naturally, Newton was determined to be the sole discoverer of calculus in the report, published as the *Commercium Epistolicum Collinii & aliorum, De Analysi promota* [anonymously written by Newton] at the end of 1712, and later published in the *Philosophical Transactions of the*

Royal Society of London in 1714. Not only that, the report determined that Leibniz was guilty of concealing his knowledge of the works of others. Leibniz was never asked by the committee for his opinion of the events.

For Leibniz, the remainder of his life did not bode well. His patron, the Elector of Hanover, became King George I of England in 1714; therefore, his supporter effectively became English and chose to favor Sir Isaac Newton, an Englishman and the celebrity of science. Leibniz would die only two years later in 1716 and his former secretary was the sole person present at the funeral.

Leibniz, unfortunately, had found himself tasked with quarreling with the most famous scientist of the time, and would lose his respect as a scientist and person by being defrauded as a plagiarist. With other evidence coming about the quarrel many years later, including Newton's placement on the committee and copies of letters by Newton and Leibniz it is now agreed that both scientists came to discover calculus on convergence and there was likely no plagiarism from either side. Of this unfortunate and malicious situation, Leibniz stated:

> It is most useful that the true origins of memorable inventions be known, especially of those that were conceived not by accident but by an effort of meditation. The use of this is not merely that history may give everyone his due and others be spurred by the expectation of similar praise, but also that the art of discovery may be promoted and its method become known through brilliant examples.
>
> (qtd. in Bardi 2006)

Newton stated: "Second inventors have no right. Whether Mr Leibniz found the Method by himself or not is not the Question ... We take the proper question to be, ... who was the first inventor of the method." He continued, "To take away the Right of the first inventor, and divide it between him and that other [the second inventor], would be an Act of Injustice." Either way, as best stated by Hall (2002), "To examine the last years of the calculus dispute does not increase one's admiration for some of the greatest of mankind."

ADVANCEMENTS IN THE NATURAL SCIENCES: FORGING THE PATH TO EVOLUTION BY NATURAL SELECTION

The Enlightenment is often considered a time of peace and prosperity, probably due to the popularity of the philosophical thinking and rationalism of the age, however, the ideals of philosophers were influenced by wars from preceding times and the wars during the Enlightenment (Starkey 2003). During the eighteenth century, for example, there were at least ten major wars that involved European countries, including The War of the Polish Succession (1733–1735), The War of Jenkins' Ear (Anglo-Spanish) (1739), the War of the Austrian Succession (1740–1748), The Seven Years War (1756–1763), The Russo-Turkish War (1768–1774), the American Revolutionary War (1775–1783), the War of the Bavarian Succession (1778–1779), The Russo-Turkish War (1787–1790), the Austro-Turkish War (1788–1790), and the French Revolution (1789–1799) (Starkey 2003).

Many of the wars during the Age of Enlightenment were not solely based on evangelism and religion, but instead based on colonialism. For example, the Seven Years War during the mid-eighteenth century was a global war between parts of Europe, North America, and Asia (Dobie 2009). Often referred to as the French and Indian War, this war resulted in the British not only ousting the French of the American Colonies, but also the occupation of India. Colonialism and war change political borders, and during the Age of Enlightenment world exploration and travel allowed greater exploration into new territories, and with that came the discovery of new species to the European natural scientists, thus requiring a new method for cataloging the diversity. The mounting discovery of the diversity of organisms coupled with the findings of fossils and advancements in geology produced a whirlwind of new ideas that would culminate into ideas of transmutation and questions of the age of the Earth, which forever changed our perception of the natural world, and the role of insects.

The field of naming and categorizing organisms, i.e., taxonomy, has probably been around as long as hominins established sophisticated means of communication (Raven et al. 1971). For instance, the ability to recognize poisonous plants, fungi, and venomous snakes would have been passed down through the generations, along with information about when it is best to hunt particular species. Although Aristotle took great strides in taxonomy by categorizing all organisms as either animals or plants, each with many subcategories, it was the advancements of technology, e.g., microscopy (discussed more in Chapters 1–3, 6, 8 in this volume), and worldwide travel during the Enlightenment that taxonomy began to adopt new forms of classification, which replaced the foundation laid by Aristotle.

Many of the early taxonomists during the Age of Enlightenment were first and foremost botanists. For example, John Ray (1627–1705) published *Methodus Plantarum Nova* in 1682, which categorized plants based on a combination of characters, a novel idea at the time. In the first of his three-volume book *Historia generalis plantarum* (1686) he provided one of the first definitions of a species, which he stated:

> ... no surer criterion for determining species has occurred to me than the distinguishing features that perpetuate themselves in propagation from seed. Thus, no matter what variations occur in the individuals or the species, if

> they spring from the seed of one and the same plant, they are accidental variations and not such as to distinguish a species ... Animals likewise that differ specifically preserve their distinct species permanently; one species never springs from the seed of another nor vice versa.
>
> (qtd. in Wilkins 2009)

Joseph Pitton de Tournefort (1656–1708) also produced important taxonomic works, such as his book *Eléments de botanique, ou Méthode pour reconnaître les Plantes* in 1694, which included information on more than nine thousand plant species. Tournefort's works were popular among botanists due to the outstanding artwork by Claude Aubriet and the fact that they further developed the concept of distinguishing between genus and species (an idea initially employed by Kaspar Bauhin, 1560–1624) (Allaby 2010), however, the relevancy of his taxonomic work is questionable. For example, Tournefort did not use some of the important taxonomic characters that were used previously by John Ray, such as characterizing plants based on the presence or absence of seeds or if they are monocots or dicots.

Although Tournefort's taxonomic work is considered to lack practicality for categorizing plants, Carl Linnaeus (1707–1778) (discussed further in Chapters 1 and 3) was greatly inspired by his book. Even as a child, Linnaeus had a deep-rooted interest in plants. In 1727 he attended Uppsala University in Sweden to pursue his passion for botany, and it was there that he wrote a thesis entitled "Praeludia Sponsaliorum Plantarum" (Plant Sexual Reproduction), which attracted the attention of the botany professor Olof Rudbech the Younger, who then offered Linnaeus to teach his lectures. Linnaeus's lectures were immensely popular, and he would become highly respected as a teacher (Blunt 1971).

At the age of twenty-five, Linnaeus embarked on his first expedition to Lapland, Sweden, a large tundrous region that was little explored. By this point Linnaeus had already determined that Tournefort's classification system was not ideal and was employing plant characters such as the numbers of stamens and pistils to categorize plants. Linnaeus's trip to Lapland had a profound effect on him—not only did his observations and naming of new plants later result in the published book *Flora Lapponica, Exhibens Plantas Per Lapponium* in 1737, but it also was on this journey that Linnaeus began to think about the applications of his naming system to those of animals. While on a roadside Linnaeus observed the lower jawbone of a horse and stated:

> If I only knew how many teeth and of what kind every animal had, how many teats and where they were placed, I should perhaps be able to work out a perfectly natural system for the arrangement of all quadrupeds.
>
> (qtd. in Blunt 1971)

In 1735 Linnaeus attended the University of Harderwijk to obtain his doctoral degree in medicine. There he completed his dissertation entitled "Dissertatio medica inauguralis in qua exhibetur hypothesis nova de febrium intermittentium causa" which discussed the possible causes of malaria, which he hypothesized involved clay-rich soils. It also was in this year that he published the first edition of *Systema Naturae*, which laid the foundation of binomial nomenclature. A book of only eleven pages in length, the first edition of *Systema Naturae* presented the new classification system that included three kingdoms: the animal kingdom, the plant kingdom, and the kingdom of stones (Blunt 1971). It is here that Linnaeus, based on anatomical features, placed humans alongside the apes, although he would have preferred otherwise, and wrote to a friend:

> It is not pleasing to me that I must place humans among the primates, but man is intimately familiar with himself. Let's not quibble over words. It will be the same to me whatever name is applied. But I desperately seek from you and from the whole world a general difference between men and simians from the principles of Natural History. I certainly know of none. If only someone might tell me one!
>
> (qtd. in Reel 2013)

In 1741 Linnaeus accepted a professor position at Uppsala University where he took responsibility for the Botanical Garden and taught botany and natural history courses. He was very popular among students, which resulted in him recruiting his seventeen apostles—students who would travel the world discovering new species and sending these to Linnaeus for him to name with his binomial nomenclature, further spreading the word of his classification system. Many of the students died prematurely, often to tropical diseases, and is why Linnaeus began only recruiting single men for these adventures, thus not to leave families without a husband or father, which was the unfortunate case with his first apostle student (Freegard 2011). In 1753 Linnaeus would publish his epoch *Species Plantarum*, which listed every species of plant known at that time, each categorized accordingly to their respective genus.

Over time, Linnaeus's cataloging, categorizing, and naming would grow into an enormous list of species, which he would use to expand each edition of *Systema Naturae*. Perhaps the most famous is the tenth edition of *Systema Naturae* in 1758, which consistently used his binomial nomenclature to classify approximately 4,400 animal species and 7,700 plant species. It is here that Linnaeus famously categorized whales with mammals and not with the fishes. Linnaeus used kingdom, class, order, genus, and species as his hierarchical system, and the common grouping of phylum and family would come later. Linnaeus would go on to inspire numerous other scientists, including Johan

Christian Fabricius (1745–1808) (discussed in Chapters 1 and 3), who was one of Linnaeus's students and would later become known as the "Linnaeus of insects" (Steffens 1842). Fabricius added approximately ten thousand more insect names to the three thousand named by Linnaeus (Tuxen 1967).

Although Fabricius is recognized for his outstanding contributions to the systematics of insects, he also was an early evolutionist, which was gaining popularity and support during the Age of Enlightenment. The growing scientific movement of transmutation (one species changing into another) was largely facilitated by advancements in geology and paleontology. Georges-Louis Leclerc, Comte de Buffon (1707–1788), for example, argued in *Histoire naturelle, générale et particulière* (consisting of thirty-six volumes published from 1749–1788) that the Earth had formed via a comet striking the sun, which created debris that formed the Earth. His radical thoughts, which employed the new physics of Newton, led Buffon to believe that the Earth was approximately seventy thousand years old. whereas most of western civilization accepted the Earth to be merely six thousand years old according to biblical scripture. Due to his estimates of the Earth's age not fitting the biblical status quo, his books were attacked by the church. Out of fear of his books becoming listed on the Index of Forbidden Books (a list of books condemned by the Roman Catholic Church), he revised the book to include a censure and apology (though likely not sincere) to the theologians who were offended by his estimates of the Earth's age (Lyon and Sloan 1981). His ideas also incorporated transmutation and that species were affected by the environment; however, he did not provide a mechanism for this change.

Estimates of the age of the Earth extending beyond those suggested by religious affiliates played a critical role in our understanding of life's natural history. The "founding father of paleotonology," Georges Cuvier (1769–1832), also provided new insight into Earth's past through the revelation of extinction, a concept previously thought impossible. His *Essay on the Theory of the Earth* (1813) established the idea that periodic catastrophic events on Earth were the primary cause of extinctions, but he was not a supporter of evolution, as a matter of fact he was firmly against it. His contemporary at the Muséum national d'Histoire naturelle (National Museum of Natural History) in France, Jean-Baptiste Lamarck (1744–1829), felt otherwise (Burkhardt Jr. 2013).

Lamarck, and his theory of inheritance of acquired characteristics, which he details in his 1809 book *Philosophie Zoologique*, is often discussed in biology courses as an example of how evolution *doesn't* work. This is unfortunate, because it overshadows the fact that Lamarck was a brilliant scientist and that he is undoubtedly the first to provide a cohesive mechanism of how evolution might operate, even though it is incorrect. Perhaps Lamarck's most celebrated work is his 1801 *Système des animaux sans vertébres*, where he classified numerous invertebrates, a term he coined (Capinera 2008), and separated the arachnids from the insects.

Cuvier could not find evidence for evolution in the fossils he studied, because of the lack of transitional species. Cuvier stated:

> I am aware that some naturalists lay prodigious stress upon the thousands which they can call into action by a dash of their pens. In such matters, however, our only way of judging as to the effects which may be produced by a long period of time is by multiplying, as it were, such as are produced by a shorter time. With this view I have endeavored to collect all the ancient documents respecting the forms of animals; and there are none equal to those furnished by the Egyptians, both in regard to their antiquity and abundance. They have not only left us representatives of animals, but even their identical bodies embalmed and preserved in the catacombs.
>
> I have examined, with the greatest attention, the engraved figures of quadrupeds and birds brought from Egypt to ancient Rome, and all these figures, one with another, have a perfect resemblance to their intended objects, such as they still are to-day.
>
> From all these established facts, there does not seem to be the smallest foundation for supposing that the new genera which I have discovered or established among extraneous fossils, such as the paleoetherium, anoplotherium, megalonyx, mastodon, pterodactylis, etc., have ever been the sources of any of our present animals, which only differ so far as they are influenced by time or climate. Even if it should prove true, which I am far from believing to be the case, that the fossil elephants, rhinoceroses, elks, and bears do not differ further from the existing species of the same genera than the present races of dogs differ among themselves, this would by no means be a sufficient reason to conclude that they were of the same species; since the races or varieties of dogs have been influenced by the trammels of domesticity, which those other animals never did, and indeed never could, experience.
>
> (qtd. in Williams 2018)

Although Lamarck and Cuvier did not agree about evolution, Cuvier made it a point to not only disagree, but to also ruin Lamarck's reputation. Cuvier's interest in ruining Lamarck's career might be based on the fact that Lamarck, although impressed by Cuvier's works, did not support his initial application to the museum, which called for an illustrator, not a paleontologist.

Lamarck would go on to point out that Cuvier's logic about transmutation was flawed and that there were not sufficient amounts of time between the embalmed organisms and their modern-day counterparts that Cuvier studied (Williams 2018). Regardless, the reputation of Cuvier as a premiere scientist overshadowed the workings of Lamarck, and because of this and his outspoken and negative views of Lamarck, he would continue to flourish as a scientist, whereas Lamarck died poor and blind.

The theory of evolution was also supported by another famous Darwin, Charles's grandfather Erasmus Darwin (1731–1802). Erasmus was a renowned doctor and surgeon during his lifetime and was even asked by King George III to become the physician to the king, which Erasmus turned down. Erasumus's true passion was the natural sciences, particularly of plants, for which he wrote numerous poems, often anthropomorphizing plant sex (King-Hele 1999). For example, in his book *The Botanic Garden* (1791), he writes:

> WEAK with nice sense, the chaste MIMOSA stands,
> From each rude touch withdraws her timid hands;
> Oft as light clouds o'erpass the Summer-glade,
> Alarm'd she trembles at the moving shade;
> And feels, alive through all her tender form,
> The whisper'd murmurs of the gathering storm;
> Shuts her sweet eye-lids to approaching night;
> And hails with freshen'd charms the rising light.
>
> Veil'd, with gay decency and modest pride,
> Slow to the mosque she moves, an eastern bride;
> There her soft vows unceasing love record,
> Queen of the bright seraglio of her Lord.—
> So sinks or rises with the changeful hour
> The liquid silver in its glassy tower.
> So turns the needle to the pole it loves,
> With fine librations quivering, as it moves.
>
> (Darwin 1791)

Erasmus Darwin's interests in animals, though, rivaled his passion about plants, and he might be one of the first to publish about the functionality of lepidopteran proboscises, writing:

> Many insects are provided with a long and pliant proboscis for the purpose of acquiring this grateful food, as a variety of bees, moths, and butterflies: but the Sphinx Convolvuli, or unicorn moth, is furnished with the most remarkable proboscis in this climate. It carries it rolled up in concentric circles under its chin, and occasionally extends it to above three inches in length. This trunk consists of joints and muscles, and seems to have more

> versatile movements than the trunk of the elephant; and near its termination is split into two capillary tubes. The excellence of this contrivance for robbing the flowers of their honey, keeps this beautiful insect fat and bulky; though it flies only in the evening, when the flowers have closed their petals, and are thence more difficult to access; and at the same time the brilliant colours of the moth contribute to its safety, by making it mistaken by the late sleeping birds for the flower it rests on.
>
> (Darwin 1791)

Erasmus Darwin would publish his famous book *Zoonomia* in 1794, which outlined his views on transmutation, including the idea of survival of the fittest (King-Hele 1999). He writes:

> Would it be too bold to imagine, that in the great length of time, since the earth began to exist, perhaps millions of ages before the commencement of the history of mankind, would it be too bold to imagine, that all warm-blooded animals have arisen from one living filament, which THE GREAT FIRST CAUSE endued with animality, with the power of acquiring new parts, attended with new propensities, directed by irritations, sensations, volitions, and associations; and thus possessing the faculty of continuing to improve by its own inherent activity, and of delivering down those improvements by generation to its posterity, world without end!
>
> (Darwin 1794)

The ideas of the Age of Enlightenment, both philosophical and scientific, would go on to forever change our views of the world. It is during this period that the old ways were replaced with the new, science became rigorous, and we began to understand our place in the universe. It also is during this time that great advancements were made in the studies of insects, which was propelled through the employment of microscopes (discussed in Chapters 1, 3, 4, 6, and 8) and facilitated our advancements in understanding the link between insects and human diseases (Chapter 2). During this period, we see that insects are not only used as symbols, finding lead roles of symbolism in our stories (Chapters 5, 6, and 7), but also are the focus of exquisite works of art (Chapter 8). Advancements during the Enlightenment would not only shape our perspective of insects, but also would show how we can apply insects to food production (Chapters 3 and 4). It is also during the Enlightenment that we first come to use the term "entomology." As stated by Charles Bonnet in his book *Contemplation de la nature* in 1764, "I have given the name insectology to that part of natural history which has insects for its object; that of entomology ... would undoubtedly have been more suitable ... but its barbarous sound terryfy'd me." Arguably, one of the most important aspects of the Enlightenment is that the progress in science and the ideas of its pioneers would come together and eventually result in the greatest discovery in the following period, evolution by natural selection, and our greatest scientist, Charles Darwin.

CHAPTER ONE

Insect Knowledge

KRISTEN E. REITER AND ROBERT HAMILTON IV

The Age of Enlightenment (1600–1820) saw unprecedented growth in the knowledge of insects. A time period of intellectualism and rationalism, scientists became more rigorous in their methods of observations and discovery, and the breakthroughs in microscopy opened an unknown world of insect structures and life histories. Scientists were less confined by traditionalism and even became more collaborative, evidenced by the explosion in scientific societies during this time period. For example, the Royal Society is an academic institution that several leading Age of Enlightenment entomologists belonged to or corresponded with its members. The Royal Society was founded and given a royal charter in 1660 as the Royal Society of London for Improving Natural Knowledge (The Royal Society 2019). Considering that the Royal Society is a scholarly society with a range of disciplines, entomologists in the Age of Enlightenment would have been exposed to a variety of scientific fields and revolutionary ideas. It is the oldest scientific academy in continuous operation in the world. Antoni van Leeuwenhoek, Charles Bonnet, William Kirby, Marcello Malphigi, René Antoine Ferchault de Réaumer, and Robert Hooke are Fellows of the Royal Society (FRS), which is discussed more in this chapter. In addition, several leading Age of Enlightenment entomologists belonged to or had correspondence with members of the Linnean Society of London. Founded in 1788, the Linnean Society of London was and still is an important forum for discussing natural history and life sciences (The Linnean Society of London 2019). The Society possessed Linnaeus's insect collection and other specimens, books, and letters. Members could examine samples. The Linnean Society received a royal charter in 1802. It is the oldest continuously operating natural history organization in the world (Gage and Stearn 1988). The Age of Enlightenment provided an environment for scientific intellectuals to explore new ideas and exchange information.

The Age of Enlightenment also represents a time period where the studies of insect specimens became an important component of entomology. The Bath white butterfly (*Pontia daplidice*), collected in England in May of 1702, is the oldest extant pinned insect (still on its original pin). The practice of pinning insect specimens likely dates to the end of the seventeenth century (Berenbaum 2017).

Reporting all of the enormous acquisition of insect knowledge during the Age of Enlightenment is far too large of scope for this chapter, however, here we present some of the important books, leading scientists, and entomologists of the time period with a brief overview of their contributions to entomology.

EARLY ENLIGHTENMENT ENTOMOLOGY

Charles Butler was an English scholar in the late sixteenth and seventeenth centuries. Like many intellectuals at the time, he had a variety of research interests. He was a logician, grammarist, author, and priest (Crane 1999). He is often referred to as "the father of English beekeeping." Much of Butler's works were based on observations of beehives he kept at the parsonage. He summarized his observations in the seminal publication *The Feminine Monarchie, or The historie of bees* in 1609. The book was originally published by Joseph Barnes and is the first full-length English language book on beekeeping. Butler significantly revised the volume in 1609, 1623, and 1634. In these volumes, Butler was the first to observe that drones were males and workers females and that bees make, rather than collect, wax. Although he was not the first to discover it, he universalized the fact that the queen ruled the hive, rather than a king as was commonly believed (Sarton 1943; Owen 2017). Butler published practical advice about the care and maintenance of beehives. His works guided beekeepers for over a hundred years until other scientists revised his works.

Insectorum sive Minimorum Animalium Theatrum (1634), or *Theatre of Insects*, authored by Thomas Muffett (sometimes Moffet), Conrad Gessner, and Thomas Penny, was the first English work solely focused on entomology (Gillis 1993; Johnston 2011). The majority of the scientific investigations in *Theatrum* were contributed by Conrad Gessner (Johnston 2011). While Gessner is most recognized for his illustrations of natural histories, he also was a physician, with many publications on medicine (Blair 2017). Upon his death in 1565, his notes and records on insects were acquired by his co-authors; however, some of his musings on the wonders of insects remain in

Theatrum (Johnston 2011). On social insects, he wrote "who can admire or praise adequately the miraculous bodies of ants and bees ... and the fact that small bodies house such complicated minds!" (Johnston 2011).

Following his death, Thomas Penny continued to build upon Gessner's work with notes of his own until his death in 1589 (Johnston 2011). Penny was among the first Englishmen to anticipate the importance of the scientific method and of collecting data before proposing theories (Johnston 2011). After Penny's death, Thomas Muffett obtained the notes of his co-authors and combined them into a publishable work; in it he incorporated the scientific writing styles of his co-authors, notably Penny, with his own creative writing (Gillis 1993; Johnston 2011). For example, although Penny explained the origins of some insects, e.g., lepidopterans, as eggs, Muffett continued to litter the text with references to traditional ideas of spontaneous generation (Johnston 2011). Muffett anthropomorphized insects, giving them traits such as guilt (beetles), joy (grasshoppers), and even divinity (social ants and bees) in order to arrange his "insect actors" into a demonstration of ethics by way of an entomological *theatrum mundi* (Johnston 2011). He wrote that insects provide a "variety of examples of vertues, whereby they may instruct their souls and teach them that otherwise they would be very wicked" (Johnston 2011). *Theatrum* also includes practical entomological advice, such as recommendations for controlling insect and other invertebrate pests (Gillis 1993). It states that weevils on corn (wheat) can be deterred by lime or garlic-infused salt water and that nematodes on figs can be controlled with the application of ashes (Gillis 1993). Unfortunately, much of the science of *Theatrum* quickly became outdated because the authors did not use any means of magnification when inspecting their insects (Johnston 2011).

MAJOR ADVANCEMENTS IN THE STUDY OF INSECTS (AND OTHER SMALL "ANIMALCULES")

Marcello Malphigi was a seventeenth-century Italian physician and Royal Society fellow who researched anatomy and physiology across multiple taxonomic divisions. Using microscopy, he made important contributions to human, plant, and animal anatomy and physiology, histology, embryology, and endocrinology. Among his many scientific contributions, Malphigi discovered and disseminated three concepts that changed entomological thinking. First, was that galls on plants do not create the insects that emerge, rather insect eggs were laid and developed inside the gall. Second, insects do not respire through lungs, but by using trachea. Third, and perhaps his most famous discovery, was the internal structure of the insect excretory system. This discovery led to the part of the insect excretory system attached to the alimentary canal, known as Malphigian tubules, to be named in his honor (Scott 1927; Arber 1942; Bradley 1985; Meli 2011).

Robert Hooke was a seventeenth-century English scientist and Royal Fellow who performed experiments and wrote on a variety of subjects. He wrote about architecture, astronomy, several areas of biology, cartography, chemistry, paleontology, and physics. He also designed and improved on existing microscopes and telescopes. Hooke is well known for developing "Hooke's Law" in chemistry and coining the term "cell" in biology (Drake 1996). Hooke's most famous work, *Micrographia*, was published in 1665. The book was ground-breaking and is still revered for its detailed drawings and drawings of cells. It was also inspirational from an entomological perspective. Many of the drawings and discourse were on insects and other invertebrates. It contained detailed drawings of anatomy and complete organisms from five different phyla. *Micrographia* set a standard for detail in invertebrate drawings (Figure 1.1) and encouraged the use of microscopic analysis of insect internal anatomy (Jervis 2013, 2014), which paved the way for future insect morphologists.

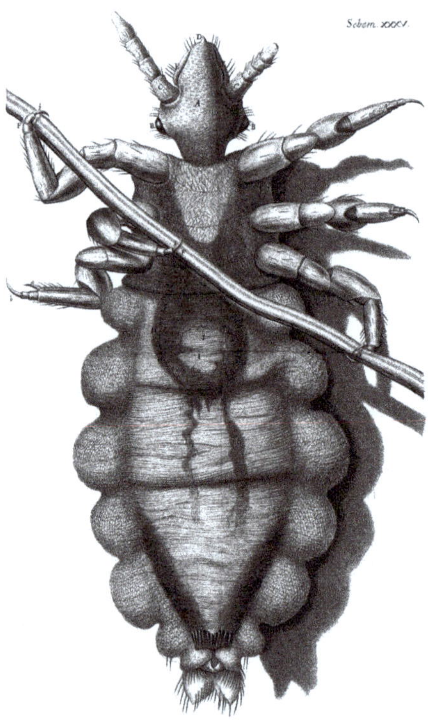

FIGURE 1.1 Schematic XXXV, engraving of a louse from Hooke's *Micrographia* in 1665. Image acquired from the National Library of Wales. Wikimedia Commons.

Jan Swammerdam was a Dutch physician active in the mid- and late seventeenth century who focused on entomological research, microscopy, and morphology. His research was innovative because he used dissection and internal anatomy to illustrate principles, whereas many of his contemporaries used descriptive language without technical terminology. Swammerdam published *Historia insectorum generalis* or *The General History of Insects* in 1669 (Figure 1.2). The book showed the imaginal discs in butterflies and challenged several erroneous widely held beliefs. He said that insects did not form from spontaneous generation. He also said that the stages of insect development (e.g., egg, larva, pupa) were not separate entities, just metamorphosis of the same creature. Swammerdam also conducted bee research and advanced the concept that the queen ruled the beehive. His most famous work *Bybel der nature*, or *Biblia Naturae*, was published in 1737, some years after his death. The printed work was noted for its drawings of the internal anatomy of several groups of insects and other invertebrates (Enenkel and Smith 2007). The book

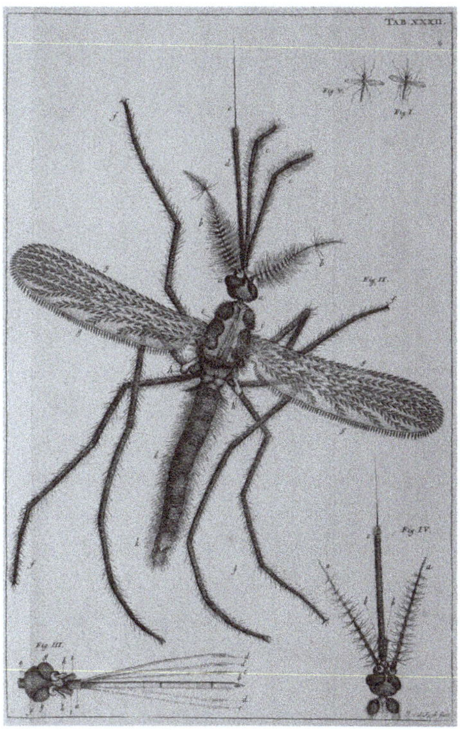

FIGURE 1.2 Illustration of a mosquito by Jan Swammerdam from *Historia insectorum generalis* in 1669. Wikimedia Commons.

also described the mechanism of muscle movement, which was commonly attributed to gases and "vital spirits" prior to his elucidation (Cobb 2002).

With his construction of high-resolution single-lens microscopes, Antoni van Leeuwenhoek was able to provide the first description of microorganisms, which he called "animalcules," in 1677 (Lane 2015). His discoveries of both protists and bacteria led to the acknowledgment of Leeuwenhoek as the father of microbiology. He described genera such as *Spirogyra*, *Euglena*, and *Vorticella*, among other protists and bacteria, and even noted subcellular structures such as cilia. Leeuwenhoek not only observed and described these animalcules, but also conducted high caliber experiments with them (Lane 2015). He understood the importance of controlling for contamination when working with these small organisms and recorded information such as dates, times, weather, and collection sources. He also studied the reproduction of cells, describing how a single cell gives rise to multiple daughters (Lane 2015). His experimental evidence supported his criticism of the theory of spontaneous generation, which would not be dismissed for another two centuries. Despite his precise records, he would not release details on his methods for microscopy (Lane 2015). Due to the inability of his peers to reproduce his findings, Leeuwenhoek and his animalcules were met with criticism and controversy. It took famed microscopist Robert Hooke three attempts to finally create a microscope that allowed him to observe Leeuwenhoek's animalcules. Hooke's support led to Leeuwenhoek's election as a Fellow of the Royal Society in 1680, where he remains the most published author in its journal to this day. In addition to his contributions to microscopy and empiricism, Leeuwenhoek was the first to suggest that aphids might be able to reproduce asexually after observing developing embryos in the ovaries of young aphids (Lane 2015).

Antonio Vallisneri was a student of Malpighi (Ratcliffe 2009). His 1696 *Dialoghi will sopra the curiosa Origine di molti Insetti* (Dialogues on the Curious Origins of Many Insects) contradicted the theory of spontaneous regeneration, claiming that the eggs of animals and the seeds of plants were the same thing. His 1704 discovery of the method of generation of duckweed silenced the insistence of a particular supporter of spontaneous generation in Rome, Father Buonanni (Ratcliffe 2009).

Charles Bonnet was a naturalist and philosopher whose scientific contributions were highly regarded by his peers (Hedges Jr. 2007). He is known for his description and research on hallucinations in psychologically normal individuals, a condition later named Charles Bonnet Syndrome. Beginning at the age of sixteen, Bonnet's studies were focused on insects. He successfully demonstrated that aphids can, as was suggested by Leeuwenhoek, reproduce parthenogenically (Hedges Jr. 2007; Davis 2012). He observed nine generations of female aphids that were all isolated at birth (Davis 2012). Bonnet, at a mere twenty years old, sent a paper on aphid parthenogenesis to the French Academy of Sciences in Paris, which resulted in him becoming the youngest member of the Academy in 1740 (Hedges Jr. 2007). He also demonstrated that caterpillars and butterflies breathe through pores, later called spiracles, which led to his 1743 membership in the Royal Society of London. At age twenty-two, Bonnet began to experience vision loss, making it difficult for him to utilize microscopy in his research; however, he continued his entomological work and contributed important studies on metamorphosis (Hedges Jr. 2007).

René Antoine Ferchault de Réaumur was a French entomologist active during the late seventeenth century through the mid-eighteenth century. He also is known for research interests in areas other than entomology, which was common during the Age of Enlightenment. De Réaumur's contributions to science allowed him to attain fellow status in the Royal Society, Académie des Sciences (French Academy of Sciences) (French Académie des Sciences 2019), and a foreign member of the Royal Swedish Academy of Sciences. His greatest contribution to entomology was elucidating the relationship between temperature and insect growth rate. De Réaumur had several other important contributions such as verifying that echinoderms could regenerate lost legs and calculating insect population growth rates (Egerton 2006b).

The theory of evolution is central to the understanding of any group of organisms, and insects are no exception. Georges-Louis Leclerc, Compte de Buffon, might have been one of the first to attempt to grapple with the complex idea of evolution. In an essay in the fourth volume of *Histoire Naturelle* (1766), "De la Dégénération des Animaux," he addressed two evolution-adjacent concepts: he discussed the production of fertile hybrids from goats and ewes, suggesting a common origin of these species, and patterns of mammal distribution in the New and Old Worlds, the differing climates and soils of which might allow for modifications that could give rise to new species (Wilkie 1956a). Although he contributed the idea that distributions have an evolutionary significance, Buffon was unable to construct a general theory about evolution (Wilkie 1956a, 1956b). Johan Christian Fabricius was a Danish zoologist who studied with Linnaeus and collaborated with Dru Drury, and named over ten thousand species of animals (Damkaer 2002; Sneli et al. 2009). He is considered one of the greatest entomologists of the eighteenth century for his transformative writings on insect systematics. His studies covered what he called "insecta," which in modern taxonomy includes invertebrate groups such as arachnids, crustaceans, and other non-insect arthropods. Fabricius is best known for his revisions to insect classification that are still widely used today. Fabricius based his insect classification on mouthparts instead of wings, which were used by Linnaeus. He also said that male genitalia also would be useful in insect classification, but he did not use this methodology himself (Grimaldi and Engel 2005). His transformative works that he published during the Age of Enlightenment include: *Systema entomologiæ* (1775), *Genera insectorum*

(1776) and *Species insectorum, exhibentes eorum differentias specificas, synonyma auctorum, loca natalia, metamorphosin, adjectis observationibus, descriptionibus* (1781), *Philosophia entomologica* (1778), *Mantissa insectorum* (1787), *Entomologia systematica emendata et aucta* (1792–1799), *Systema eleuthatorum* (1801), *Systema rhyngotorum* (1803), *Systema piezatorum* (1804), *Systema antliatorum* (1805), and *Systema glossatorum* (1807). These collective works permanently changed insect classification. He also wrote about evolution, but from a Lamarckian perspective.

NATURAL HISTORY IN ENTOMOLOGY: DESCRIBING INSECTS ACROSS THE GLOBE

John Ray and Francis Willughby were leaders in the field of insect natural history in the latter half of the seventeenth century and early 1700s (Ogilvie 2012). Willughby and Ray shared an interest in metamorphosis and diversity, shown in Ray's 1705 *Methodus insectorum* and 1710 *Historia insectorum*, both containing material from Willughby. They described diverse insects from their personal collections as well as those from London and Oxford and developed the novel use of metamorphosis in their classifications (Ogilvie 2012). Ray's *Historia insectorum* was particularly viewed by his contemporaries as a significant contribution to the emerging field of entomology.

William Dandridge Peck was the first Harvard professor of natural history and Fellow of the American Academy of Arts and Sciences during the mid-eighteenth century to the early nineteenth century. As a naturalist he conducted surveys of several taxonomic groups yet specialized in botany and secondarily entomology. At the time most entomological and other natural history research was conducted in Europe. His work is significant because his publications were among the first zoological works published in America. Peck's collections of North American flora, fauna, and entomofauna increased the known biological diversity (The Museum of Comparative Zoology, Harvard University 2019). He is considered to be the first native-born American entomologist.

Carl Alexander Clerck was an eighteenth-century Swedish entomologist and arachnologist. He collected and classified spiders in the mid-eighteenth century. His seminal work was *Svenska Spindlar* (Swedish spiders), or *Aranei Svecici*, in 1757 (Clerck 1757). It was the first work to consistently use binomial nomenclature as is common today. He also defined "spider" (*sensu strictu*) with anatomical traits still used today to define the taxon. His spider nomenclature and classification were accepted by Linnaeus with minor revision and included in *Systema Naturae*. He also collected and identified butterfly specimens but died before he could publish about them (Mikkola 1985). Clerck was elected a fellow of the Royal Society of Sciences in Uppsala (Kungl.

Vetenskaps-Societeten i Uppsala) in 1756 and of the Royal Swedish Academy of Sciences (Kungl. Vetenskapsakademien) in 1764.

The tenth edition of Carolus Linnaeus's *Systema Naturae* is unanimously regarded as the origination of the rule of priority in giving animals scientific names (Calman 1939). It was the first edition of *Systema Naturae* to use binomial nomenclature consistently throughout its descriptions (Calman 1939). *Systema Naturae* was intended to not only list species and their characteristics, but to also summarize all known information of each species; many entries include details on biology, ecology, and geographical distribution (Calman 1939). Published in 1758, it was the last edition of *Systema Naturae* to appear during Linnaeus's lifetime (Calman 1939).

Moses Harris began collecting insects from the age of twelve, a practice he inherited from his uncle (Weiss 1926). In the following two decades he began a career as an entomologist, collecting and illustrating insects, especially Lepidoptera (Weiss 1926). This work culminated in *The Aurelian or Natural History of English Insects; namely, Moths and Butterflies, together with the Plants on which they Feed* (1766). *The Aurelian* includes information on insect collections and lepidopteran life histories; descriptions of the appearance and markings of adult butterflies and moths, host plant information, how to use collecting nets, and how to prepare and spread collected insects (Weiss 1926). It also features an illustration of Harris collecting insects with a net. Harris later authored entomological publications such as *The English Lepidoptera, or the Aurelian's Pocket Companion, containing a Catalogue of upward of four hundred Moths and Butterflies, &c* (1775) and *An Exposition of English Insects* (1776), which described Diptera, Neuroptera, and Hymenoptera, revealing Harris as more than only a lepidopterist (Weiss 1926). Harris described several new insect species throughout his work (Cockerell 1922). In his 1767 *An Essay preceding a Supplement to the Aurelian wherein are considered the Tendons and Membranes of the wings of Butterflies*, Harris pioneered the idea of classifying insects according to wing venation, a trait that remains useful in identification and classification to this day (Cockerell 1922; Weiss 1926). Harris also created the majority of the colored engravings in Dru Drury's *Illustrations of Natural History* (Cockerell 1922; Weiss 1926).

One of the fathers of English entomology, Dru Drury was a naturalist and collector who described a large number of insects (Cockerell 1922). He published the three-volume *Illustrations of natural history, wherein are exhibited figures of exotic insects* (1770–1782) with illustrations by Moses Harris. These volumes contain illustrations and descriptions of over a thousand foreign species of insects sorted by genera with brief descriptions of their natural histories (Cockerell 1922). Although the text of *Illustrations of natural history* did not include scientific names, the index held binomial names for each insect following the method of Linnaeus, a naturalist who Drury greatly respected. Drury was the first Englishman to thoroughly adopt the Linnaean method of naming species (Cockerell 1922). Drury not only conducted his own entomological research, but also encouraged others to study insects. In a letter to Thomas James of New York, he urged James to study the immature forms of dragonflies (Cockerell 1922). He also was known to persuade Fabricius, a student of Linnaeus and later entomologist, to study insects (Cockerell 1922). This encouragement might have come at a disadvantage to Drury, however: he left his collections open for observation by his students, such as Pieter Cramer, who from time to time would describe new species without Drury's consent and without the use of his suggested names (Cockerell 1922).

Edward Donovan began his studies of natural history investigating plants, publishing *The botanical review, or the beauties of flora* (1789–1790) (Gibson 2017). It included accurate illustrations of true-to-size and magnified foreign species of plants—these high-quality scientifically accurate figures were the selling point of the work (Gibson 2017). *The botanical review* also included practical information on the cultivation of the illustrated plant species (Gibson 2017). Donovan would become well known for his mix of aesthetics and empiricism, first exemplified by *The botanical review*. His books were written for a broad audience, appealing to serious scientists and amateur naturalists alike: he published both an instruction manual for beginner collectors, *Instructions for collecting and preserving various subjects of natural history; as animals, birds, reptiles, shells, corals, plants, &c* (1794), and a detailed sixteen-volume work on the insects of Britain, *The Natural History of British Insects* (1792–1813), around the same time (Gibson 2017). *Instructions* provided extensive information of the methods of studying natural history rather than the natural histories of the animals themselves in order to assist amateurs in entering the field. It included details on how to collect and handle specimens and stressed the importance of examining live specimens (Gibson 2017). On the other hand, *The Natural History of British Insects* appealed to seasoned naturalists and included Latin descriptions of insect characters (Gibson 2017). As was the case in *The botanical review*, Donovan's illustrations (thirty-five per volume) were the focal point of *British Insects*, many highlighting the life cycles of insects. The process of insect development, from egg to larva to pupa to adult, was of great philosophical interest at the time; Donovan took advantage of this and included metamorphosis in *British Insects* to rouse interest in his work (Gibson 2017). The volumes have been described as an intermediary between the more art-centric works of the eighteenth century and the empirical works of the nineteenth century. Following his own advice in *Instructions* and the precedent set by Maria Sibylla Merian, Donovan drew from live specimens in order to create the most credible illustrations and was therefore regarded for the accuracy of his work (Gibson 2017). The use of live insects for illustrations and descriptions represented a major advance in entomology, allowing for more accuracy in not only illustrations but also in later identification and classification (Gibson 2017). Donovan went on to publish a number of books on foreign entomology: *An Epitome of the Natural History of the Insects of China* (1798), *An Epitome of the Natural History of Insects of India, and the Islands in the Indian Seas* (1800) (Figure 1.3), and *An Epitome of the Natural History of Insects of New Holland, New Zealand, New Guinea, Otaheite, and Other Islands in the Indian, Southern and Pacific Oceans* (1805) (Gibson 2017). Through his publications and later his museum, the London Museum and Institute of Natural History, he made his collections available to the public. He became a fellow of the Linnaean Society in 1799 with the support of six existing members, rather than the usual three, due to the many volumes on natural history he had already published (Gibson 2017).

FIGURE 1.3 Two species of *Papilio* butterflies illustrated by Donovan (pl. 20) in his book *An Epitome of the Natural History of Insects of India, and the Islands in the Indian Seas*. Wikimedia Commons.

Georg Wolfgang Franz Panzer was a German physician who also studied botany and entomology in the mid-eighteenth century to the early nineteenth century. He is best known for having a large and diverse insect collection and publishing *Faunae insectorum germanicae initia* (*Elements of the Insect Fauna of Germany*) in 1796 and 1813. It contained over 2,600 colored plates of life-sized insects illustrated by Jacob Sturm and was the first major catalog of German insects and beneficial to insect taxonomy beyond the region. Jacob Sturm also illustrated Franz's *Deutschlands Insectenfaune et Kritische Revision der Insectenfaune Deutschlands.*

Jacob Hübner was an etcher, artist, and naturalist who contributed numerous publications and generic names to the field of entomology (Hemming 1937). Born in 1761, Hübner grew up in Augsburg, Germany. As a child, his artistic abilities stood out. He apprenticed as an etcher and became successful in the profession. In his early twenties Hübner found an interest in natural history. By 1784 he began attempts to depict butterflies and moths in his engravings. The following year he published his first work on Lepidoptera, *Abbildungen und Beschreibingen noch nicht abgebildeter und noch unbeschriebener Schmetterlinge* (1785), which described new species. Hübner's interests soon moved to the early life stages of lepidopterans, publishing *Beiträge zur Geschichte der Schmetterlinge* in 1786. This illustrated work had a great impact on the study of Lepidoptera as very little was known about their early developmental stages at the time. In order to take a position as designer at a cotton factory, Hübner moved to Ukraine for three years. During his time there, he collected many butterflies and moths. This opportunity to collect new species fueled Hübner's interest in collections. His collecting efforts resulted in the compilation of a list of Ukrainian Lepidoptera, *Verzeichniss ukrainischer Schmetterlinge in den Gegenden un Niemierow*, which was never published, but is held at the Royal Entomological Society of London's library.

Hübner's move to Ukraine also allowed him to meet with and study the collections of foreign entomologists. The notes he took while studying their specimens and libraries would become important in many of Hübner's later works (Hemming 1937). Hübner returned to Augsburg to continue working as a designer, but his interest in natural history remained intact. He briefly dabbled in ornithology, publishing a work on both birds and insects, *Sammlung auserlesener Vögel und Schmitterlinge* (1793), before returning his focus to Lepidoptera. He began working on a systematic catalog of Lepidoptera, which resulted in the publication of *Der Schmitterlinge, Lepidoptera Linnei, Europäisches Heer* (1790–1793). His interests would return to cataloging the world's butterflies and moths over twenty years later. In the meantime, he devoted most of his efforts to creating his earliest monumental work, *Geschichte europäischer Schmetterlinge*. This work illustrated the early life stages of European Lepidoptera. He collaborated with artists Hörmann and Schiffermüller who provided painted figures for *Geschichte*. The eighty illustrations contributed by Hörmann were considered exceptionally exquisite, and the sixty drawn by Schiffermüller were not as beautiful but depicted the insects with both skill and accuracy. Four hundred and fifty-seven plates were published for *Geschichte* by the time of Hübner's death in 1826, with over five hundred total plates published by its last volume.

Hübner wanted to publish a companion to *Geschichte*, which would feature illustrations of all of Europe's adult Lepidoptera. At the time there existed only one illustrated work on adult butterflies and moths, but its plates were crude, and it did not include Microlepidoptera. Thus, *Sammlung europäischer Schmetterlinge* began publication in 1796 with sections dedicated to pyralids and sphyngids, with later illustrations of other groups published over time. Over the next ten years, Hübner focused on the publication of plates for *Geschichte* and *Sammlung*. During this process it became apparent to Hübner that to rationally classify the order, Lepidoptera should be further subdivided. He expressed this idea in his 1805 *Tentamen*, which outlined his proposed system for classification. His interest in classifying Lepidoptera grew and he published over 1,600 generic names from 1806 until his death. He applied the nomenclature suggested in his *Tentamen* to the exotic lepidopterans he described in *Sammlung exotischer Schmetterlinge* (1806).

Hübner realized that in order to study any group of animals, species from every region should be examined. He continued publishing plates for *Sammlung exotischer* over the next twenty-two years, resulting in 434 plates published by his death. In 1808, Hübner partially published a companion to his work on foreign Lepidoptera, *Erste Zuträge zur Sammlung exotischer Schmetterlinge*. Later he published the work in its entirety under the title *Zuträge zur Sammlung exotischer Schmettlinge*. By his death, plates for the first three volumes and text for the first two volumes of *Zuträge* were published. The remaining plates and text were published after Hübner's death. From 1808 onwards, he focused on publishing plates for all four of his major works, an impressive feat as he continued his work at the cotton factory. As wartime violence subsided, Hübner was able to focus more of his efforts on his entomological studies (Hemming 1937). This resulted in the 1816 publication of *Verzeichniss bekannter Schmettlinge*, a systematic catalog of the world's Lepidoptera, achieving the goal he had set twenty-five years prior. *Verzeichniss* was the first attempt since the time of Linnaeus to publish a comprehensive catalog of species. It included Hübner's systems for classifying and naming Lepidoptera, with hundreds of generic names first appearing in this work. *Verzeichniss* is considered Hübner's greatest work and was his last major publication. Following the death of his wife in 1820, Hübner published few plates and no major works. In 1825 he published his final work, a description of the collection of a friend: *Catalogue des Lépidoptères qui composent la Collection de feu Mr. Franck*. He died the following year at the age of sixty-five.

LATE ENLIGHTENMENT ENTOMOLOGY

William Kirby was an English entomologist in the late eighteenth and early nineteenth centuries. While many of his works were produced at the end of the Age of Enlightenment and beyond, his contributions during this period laid the groundwork for future works in entomology. He is often referred to as the "Founder of Entomology." William Kirby also was a priest and referred to as a "parson-naturalist," which is someone who saw their natural science research as an extension of their religious studies (Armstrong 2000). It was not so uncommon at the time for entomologists, or other scientists and philosophers, to also practice Christianity (Kors 2002). He was an original member of the Linnean Society of London (1788) and a Fellow of the Royal Society (Armstrong 2000). Kirby published a number of volumes on entomology, but one of his most renowned works is *Monographia Apum Angliae*, about bees. William Kirby carried on the tradition of much earlier apiculturists such as Charles Butler.

The most eminent of early American taxonomic zoologists, Thomas Say is regarded as the father of a number of fields including American zoology, American conchology, American entomology, and American descriptive entomology (Weiss and Ziegler 1931). The majority of Say's work focused on insects and mollusks, but he also studied mammalogy, herpetology, and ornithology (Weiss and Ziegler 1931; Brodhead 1993). Over his entomological career, Say described over a thousand species of Coleoptera, over two hundred dipterans, approximately one hundred species of Hemiptera, and hundreds of other insects (Weiss and Ziegler 1931). His famed *American Entomologist* (three volumes: 1824, 1825, 1828) was intended to display the insects of the United States and features colored engravings (Weiss and Ziegler 1931; Brodhead 1993). He also published six volumes of *American Conchology*, an attempt to cover all North American mollusks, before his death (Weiss and Ziegler 1931; Brodhead 1993). In addition, Say published a number of descriptive papers on insects in the *Journal of the Academy of the Natural Sciences of Philadelphia*, *Transactions of the American Philosophical Society*, and other journals throughout his life as well as posthumously (Weiss and Ziegler 1931). He was a founding member of the Academy of Natural Sciences in Philadelphia and served as its journal's editor (Brodhead 1993). Say also was able to travel abroad for his studies, including field work with Stephen G. Long's expeditions and collecting trips to Florida and Mexico (Weiss and Ziegler 1931; Brodhead 1993).

In conclusion, the Age of Enlightenment saw a number of advancements in entomology. In particular, the advancements in microscopy during this period provided a technology that allowed scientists to begin making observations and illustrations of insect structures, thus the field of insect morphology emerged. The knowledge, though, of insects reached beyond insect morphology and included important contributions to insect physiology (e.g., metamorphosis, internal structures and their functions) and ecology (e.g., insect-plant relationships). Giving these findings, one could argue that the Age of Enlightenment is the age of entomology.

CHAPTER TWO

Insects and Disease

WILL K. REEVES

The vector-borne diseases transmitted by insects and other arthropods to humans and domestic animals are among the most significant sources of suffering, death, and economic loss across the globe. While outbreaks of vector-borne diseases, such as the 2016 Zika virus epidemic claim modern news headlines, they pale in comparison to the fatal plagues and pandemics of the seventeenth–nineteenth centuries. Modern medicine, hygiene, vector-control, and sanitation have alleviated many of these threats, especially in highly developed countries, but during the Age of Enlightenment (1600–1820) the causes of these diseases were almost never known or adequately prevented. The modern concept of drugs that are used to cure disease or alleviate symptoms were essentially non-existent beyond basic herbal and mineral remedies that were often of variable unstandardized concentration and ingredients. During the Age of Enlightenment diseases were still largely believed to be caused by magic, spirits, and witchcraft. There are numerous accounts throughout the period of the torturing and killing of innocents accused of witchcraft or spreading of plagues (Garrison 1926). As medical and epidemiologic knowledge progressed, the increased commerce, war, and travelers were understood to bring disease, such as plague or yellow fever. Prior to the Age of Enlightenment and based on this often colloquial knowledge there were successful naval quarantine stations, which prevented the introduction of disease (Garrison 1926; Adler and Wills 2003). The term "Yellow Jack" was coined sometime during this age to refer to both the naval quarantine flag or "jack" and to the disease yellow fever, which it often signified.

The history of medicine and human disease includes inference of causes of epidemics. Until the recent application of forensic and diagnostic tools in modern medicine, many historic epidemics simply cannot be reliably identified (e.g., Gilbert et al. 2004). The further they are in the past and the more generations removed, the more likely they are poorly remembered, inaccurately described, or possibly attributed inaccurately. Both cryptic or vividly described epidemics might be mythological or fiction later taken as fact. These include many near legendary events from ancient history that are reinterpreted often with scant facts as recently discovered diseases (e.g., Trevisanato 2007). While some of the more devastating vector-borne diseases have distinct presentations and are possible to identify, the death rates of older outbreaks are hard to accurately know (Outram 2001). Even well-known diseases, such as malaria, can be difficult to identify based only on signs and symptoms, especially when they are recorded by different medical practitioners, lay people, or church death records. Some epidemics that cause birth defects and are cryptic, or longer-term diseases, such as heart failure from progressive endocarditis, can be obscure and difficult to identify even with recent historic records (Rasmussen et al. 2016). In some cases, even the names of diseases have changed, for example chikungunya and dengue changed names over time, where the older outbreaks of chikungunya in the Americas during the Age of Enlightenment were called dengue (Halstead 2015).

Scientific developments in human and veterinary medicine led to the steady albeit slow discovery of the causal associations or agents of numerous diseases and even basic human anatomy (Cox 2002). Bacon, a British lawyer and philosopher, published what is generally accepted as the scientific method in 1620 in a book *Novum Organum Scientiarum* and advocated testing of natural events (Pesic 1999). Prior to the Age of Enlightenment much of human anatomy and physiology were unknown. In fact, the study of human cadavers was largely illegal and forbidden, but organ systems were increasingly described following the 1600s. For example, the circulatory system was described in relatively modern terms by Harvey in the book *De Motu Cordis* (Harvey 1628), and while a far older description by Ibn an-Nafis possibly deserves credit for the discovery, this knowledge was largely lost. Microscopic life was first seen during the Age of Enlightenment. The first microscopic observations of protozoa were published by van Leewenhoeck (1677); who also apparently documented the first-known parasitic protozoa from his own feces. An actual scientific demonstration that insects could support the development and transmission of a parasite of humans or domestic animals was not published until Melnikoff (1869) described the development of larval tapeworms in dog chewing lice. Mosquito-borne transmission of an obligate human parasite was later described by Manson (1883), who published on filariasis in China, but his hypothesis on the transmission route was flawed. Some of the most dreaded diseases, such as plague and yellow fever, etched their names in human history during the Age of Enlightenment and have retained their infamy in modern times despite the considerable drop in mortality and successful preventions or treatments.

The origins and epidemiology of many human diseases are often tied to urbanization, long-term associations with domestic animals, climate, and cultural or trade ties with other regions (Morand et al. 2014). Many of the most serious diseases of humans, such as measles or most recently Covid-19, probably originated from diseases of domestic or food animals that adapted to human hosts (Furuse et al. 2010; Andersen et al. 2020). People from cultures without close contact to domestic animals, nomadic people without livestock, and those who lived in isolated islands or on historically isolated continents usually have fewer serious endemic diseases. Unfortunately for these peoples, they have suffered cataclysmic death rates after encountering foreign peoples and their diseases. Diseases such as malaria, smallpox, measles, epidemic typhus, various herpes, tuberculosis, hookworm disease, and influenza killed millions of native peoples across the world after they encountered Europeans, Africans, and some Asians (Thornton 1990). This was one of the lasting true and unfortunate legacies of the Age of Enlightenment.

Specifically, four vector-borne diseases—plague, epidemic typhus, yellow fever, and malaria—were among the most significant agents of death, cultural change, military failure, and economic loss in the Age of Enlightenment. The native populations of Australia, Oceania, and the Americas probably suffered more losses from vector-borne diseases and other diseases post European contact than they had from all previous causes of death prior to that event (Crosby 1976; Thornton 1990). Most of these people lacked genetic resistance, such as sickle cell trait or Duffy variations that protect against malaria (Miller et al. 1976), or close cultural contact with livestock and poultry. There are a few possible exceptions of catastrophic diseases spread to Europe from new colonies, such as syphilis and possibly epidemic typhus, which could have spread from the New World to Europe and Asia (Burns et al. 2014).

History, as we know it, is in part a fabrication or propaganda created and written by conquerors, survivors, religions, and governments. This is especially true in the era of European expansion. Large swaths of documents and oral histories were lost in political purges and wars. The stories remembered and retold are those of the survivors and of those with the political and military power, education, and money. The true cost to human life from vector-borne diseases are practically impossible to account for in places such as Australia, Africa, and the Americas where European conquest, slavery, and religious powers purged local customs, language, and history along with a great majority of the endemic population (Thornton 1990).

While not often considered by medical historians as a cause of human suffering, war, economic collapse, or death, the diseases of domestic animals, agricultural crops, and even domesticated insects, such as silkworms and honey bees, historically had major influences on people and the welfare of countries. For example, epizootics of microsporidiosis in silkworms caused a near total collapse of the early European silk industry (Bhat et al. 2009). In terms of agricultural damage during war, the Hessian fly (*Mayetiola destructor*) caused crop losses during the American Revolutionary War in wheat, barley, and rye, and even the common name was taken as a racially inflammatory slur directed at foreign mercenaries (Pauly 2002). African trypanosomiasis has long prevented the movement of non-resistant livestock through the tsetse belts of Africa. The resulting and ongoing losses from livestock deaths and poor fitness historically prevented people from the tsetse belt from thriving. The indirect loss of human life and income from diseases of agriculture are not considered in this chapter, not because they are irrelevant, but to limit the scope to a more manageable level. Insects, such as wasps, also directly kill by envenomation and associated anaphylaxis, and while this is significant (Voss et al. 2016) it is not accounted for here either. Essentially every aspect of life in the Age of Enlightenment was touched by arthropod-borne disease.

Arthropod-borne diseases such as plague, typhus, malaria, and to a lesser extent the filarial nematodes spread to every inhabited continent during the Age of Enlightenment. European expansion, colonialism, slave trade, and industry was one of the major drivers of the expansion and introduction of disease. Each continent will be treated separately, starting with Europe.

EUROPE

European and Asian trade, travel, and wars were the major drivers of world change. Two insect-transmitted diseases—plague and epidemic louse-borne typhus—probably drove the arc of European and world history by killing millions of European soldiers, sailors, and civilians across the globe. Possibly no diseases were so influential to history than this pair in combination. They were spread within Europe and to their colonies. Both are transmitted by ectoparasites and require urbanization for sustained epidemics.

Vectors such as the Oriental rat flea (*Xenopsylla cheopis*), human flea (*Pulex irritans*), and body louse (*Pediculus humanus*) were well established across Europe prior to the seventeenth century. *Pulex irritans* was apparently introduced to Europe by the sixteenth century from the Americas by Norse sailors (Buckland and Saddler 1989). *Xenopsylla cheopis* is a parasite of rodents and probably arrived with rats in ships or cargo from Africa or Asia. Human body lice (*P. humanus*) were common on all walks of life from royalty to peasants and probably coevolved with humans. The body louse was already established worldwide. Unfortunately for Europeans, these lice and fleas were vectors of some of the most lethal diseases of the Age of Enlightenment.

Possibly the most iconic of all vector-borne diseases, plague, caused by *Y. pestis*, was embedded in European memory and history after the great Black Death of the sixteenth century (Raoult et al. 2000; Drancourt et al. 2007). It classically occurs after epizootics in peridomestic rodents. Plague is a zoonotic disease spread from rats to humans by rodent fleas, especially the flea (*X. cheopis*). There was likely some transmission from human to human with pneumonic plague, and by the human flea (*P. irritans*), which is not as successful a vector as *X. cheopis* (Lorange et al. 2005). Plague remained one of the most relevant diseases of Europe during the Age of Enlightenment causing millions of deaths (Link 2005). Records, while not totally accurate, report monumental plague deaths until around 1720. As a disease it was well known and presumably identifiable to Europeans as it had killed nearly one-third of the population in previous centuries (Gage and Kosoy 2005).

Epidemics of plague broke out throughout all of Europe in sporadic events throughout the sixteenth–nineteenth century. A number of these plague epidemics devastated European cities, such as London in 1665–1666, Great Plague of Seville (1647–1652), and Great Plague of Vienna (1679) (Sloan 1973; Greenberg 1997). There were numerous deaths throughout Italy and southern Europe during the period in smaller epidemics and isolated fatal cases (Alfani 2010). Over a million of the people of France died in one large plague outbreak (Parker 2001). The last major plague epidemic in Europe ended around 1720 in Marseilles, France. Plague outbreaks in mainland Europe killed numerous soldiers during massive extended wars, such as the Germanic Thirty Years War, where the majority of military deaths probably were caused by plague and typhus. Plague outbreaks in the twenty years of the Great Northern wars caused thousands of deaths in both European and Russian troops. During these ongoing outbreaks and the enzootic periods, European ships spread plague across the world leading to a pandemic by spreading the disease to colonies as far flung as North America, Asia, and Africa.

Epidemic louse-borne typhus emerged in Europe as a major cause of death in the sixteenth century. Based on the nature of the disease, which apparently caused a new and massive epidemic, there is a possibility that the causative agent, *Rickettsia prowazekii*, was recently introduced into a naive European population (Emmanouil et al. 2016) or evolved from a less virulent endemic relative. Whether this is true remains unknown; however, recent molecular verification of the agent from human remains verified outbreaks of the disease killed millions of Europeans between the sixteenth to twentieth centuries (Raoult et al. 2006). Reports of ancient European and Asian outbreaks of this disease are hard to verify and there is a hypothesis that the causative agent, *R. prowazekii*, evolved recently in Europe from a related *Rickettsia* near the end of the sixteenth century (Emmanouil et al. 2016). An alternative hypothesis is that it originated in the Americas and was introduced to Europe after

Europeans exploited the Americas (Emmanouil et al. 2016). The only known animal reservoir for *R. prowazekii* is the northern flying squirrel (*Glaucomys sabrinus*) in North America, which gives further support that the pathogen is a zoonosis from the Americas (McDade 1987).

Epidemic typhus probably ended or changed the course of most European wars and festered as a disease in European and nearby Asian cities that traded with them. This is in part due to the fact that the infectious agent can remain relatively dormant in people who survived the initial disease and re-emerge later as Brill-Zinsser disease. The first major European war in the seventeenth century to see mass mortality from epidemic typhus was the Thirty Years War of 1618–1648 primarily in Germany (Outram 2001). The military actions of the Thirty Years War were responsible for incredible suffering across central Europe. Estimates of deaths from the conflict are hard to verify, but range from six to ten million people. Of those deaths, less than four hundred thousand were reported from combat. Epidemic typhus and plague were believed to be the leading causes of death in both civilians and soldiers (Outram 2001). During the English Civil War, the armies of Charles I were unable to attack London because of an epidemic of typhus. The war ended with Cromwell becoming Lord Protector and the British Royalty fleeing to continental Europe (Gaunt 2014). French soldiers also returned from the Napoleonic campaign in Russia after suffering catastrophic losses from typhus; and soldiers from the Grand Army that survived often returned with body lice and some spread infection to start local outbreaks in Europe (Raoult et al. 2006).

Epidemic louse-borne typhus was probably accompanied by the less fatal disease, now named "trench fever" caused by *Bartonella quintana*. This bacterium was present in Europe and was recently discovered in four-thousand-year-old human teeth from European graves (Drancourt et al. 2005). The reservoir for *B. quintana* is humans and it is spread primarily by body lice (Drancourt et al. 2005). The disease, trench fever, was not discovered and named until the First World War when it caused large outbreaks in the European trench warfare front, but like many ailments it almost certainly caused suffering and death prior to modern nomenclature and discovery. While trench fever is far less fatal than epidemic typhus, it can be a serious and fatal disease. Several other chronic conditions, such as endocarditis, are caused by *Bartonella* infection and in the pre-antibiotic era would have been fatal. Molecular studies of materials recovered from the Napoleonic Wars clearly indicate that *B. quintana* was present in the French soldiers (Raoult et al. 2006).

Louse-borne relapsing fever, from *Borrelia recurrentis*, is a third highly pathogenic louse-borne disease capable of causing massive epidemics with high mortality rates. It is transmitted by body lice and could have killed troops during European wars. There are fewer good records on louse-borne relapsing fever across Europe, but potential epidemics occurred across Ireland and England during the Age of Enlightenment (Raoult and Roux 1999; Cutler 2006). Louse-borne relapsing fever currently has been almost eliminated from much of the world, but even with modern antibiotics it still causes up to 4 percent mortality and hemorrhagic syndromes (Lescot et al. 2008). Unlike *R. prowazekii*, which might have evolved in the Americas, *Borrelia recurrentis* is likely native to the Old World.

The ectoparasite-borne diseases were not the only scourges of Europe. Europeans also transported vivax malaria, caused by *Plasmodium vivax*, from Africa or Asia and epidemics spread across the warmer parts of the continent far north to cool temperate regions, such as the United Kingdom, Switzerland, and Holland, which possibly led to high death rates in certain naïve communities (Kuhn et al. 2003; Adams et al. 2011). However, malaria as a distinct disease from other recurring fevers was not fully understood at the time. Malaria-like illnesses with recurring fevers and chills were common and reviewing historical literature requires inferences of malarial cause from references of "the ague" or periodic fevers such as those described in parts of the UK by Caliban in Shakespeare's *Tempest* (Reiter 2000).

From 1700–1900 the people living in the coastal swamps of the UK were more likely to die than those in other regions of the UK and malaria probably played a role in those deaths (Hutchinson and Lindsay 2006). Early investigations implied that malaria was the primary cause of death, but close investigation of the seasonality and demographics of those deaths directly contradict the full relevance of malaria and imply other agents were also involved (Hutchinson and Lindsay 2006). Malaria was present in Europe long before the Age of Enlightenment, but the slave trade in Africa and the exploitation and wars with Asia brought most of Europe into continual contact with this disease.

The world's first truly successful antimalarial drug was derived from South American cinchona, which was brought to Europe in the seventeenth century. Long before that was discovered, Europeans had used various herbal remedies to treat the symptoms of malaria (Reiter 2000; Adams et al. 2011). Most of these probably had no true beneficial effects or even were toxic, but some herbal remedies do show antipyretic properties or even antiprotozoal activity in the laboratory.

Various endemic zoonotic diseases must have been present in Europe through the period. These include the parasitic protozoa (*Leishmania* spp.) in the Mediterranean. Europeans were aware of the disease in Asia and the clinical progression was described by a Scottish physician Russell (Steverding 2017). The Spanish reported disfiguring facial damage among their South American plantations that agree with the presentation of leishmaniasis. There were no major epidemics of filariasis in Europe despite the frequent contact with Africa and the almost certain established populations of mosquitoes in the genus *Culex*, such as *Culex pipiens*, which could serve as vectors. Arboviruses, such as sandfly fever viruses (*Phlebovirus*), louping ill (*Flavivirus*), Sindbis virus (*Alphavirus*), and others almost certainly caused unnamed diseases throughout the history of Europe. Unfortunately, for medical historians there was no way to identify these ailments or their causes in the scattered written records.

AFRICA

Humans (*Homo sapiens*) probably evolved in Africa. Africa is largely tropical, but the range of climates across the continent extends from tropical rainforests to deserts to high elevation alpine zones. Africa supported human civilization, rural and urban centers, and large empires for hundreds of thousands of years and our primate ancestors for millions of years before that (Scally and Durbin 2012). Parasites and pathogens have evolved with us and some of the oldest records of obligate parasites of humans come from Africa (Cox 2002). Many domestic animals and even crops originated in Africa, as did the pathogens that evolved with them (Cockburn 1971).

Africa supported a wide range of political entities during the Age of Enlightenment, which ranged from tribes to kingdoms and empires, which rivaled much of what was seen in Europe (de Graft-Johnson 1957). Many of these African groups worked, fought, and traded with European and Asian powers. They both participated in and suffered from the slave trade as a multinational affair that spread disease and misery across the globe. European colonial powers and the slave trade had direct impacts on the spread of diseases to and from Africa. Perhaps some of the most significant diseases to be spread across the Americas, Asia, and Europe arrived directly or indirectly from Africa, such as malaria and schistosomiasis, along with their vectors including several species of mosquitoes, ticks, and snails. As with Europe, both louse-borne typhus and plague were serious threats in Africa. Plague was established in African cities, especially those with large urban ports. There were thousands of plague deaths in Egypt in the 1800s. European troops in invasion, suppression, or policing forces often suffered serious losses from diseases while trying to conquer or control parts of Africa. Ships from Africa or southern Europe transported the urban mosquito (*Aedes aegypti*) across the Atlantic and probably led to the rapid evolution of urban forms within this species, which had to adapt to survive on sailing ships where it bred in cisterns and later port cities (Powell and Tabachnick 2013). These mosquitoes escaped the ships and infested port cities across the Americas, including those of the large inland river systems, such as the Mississippi, and even re-infested African ports (Powell and Tabachnick 2013).

European missionaries traveled across Africa and strove to change the culture, religion, economy, and to modernize the continent in the view of Europeans. Many taught that traditional practices and medicine were flawed and worked to teach European values, culture, language, math, medicine, and civilization to the African people. Being that European medicine was not all that advanced, it was hardly a major benefit to many Africans. In addition, some changes brought about by missionaries caused the spread and growth of endemic diseases. For example, in regions with the vector-borne parasitic disease onchocerciasis, traditional villages were isolated from rivers. This must have seemed like the planning of uncivilized people to missionaries, because they encouraged the relocation of villages. Once people were living near the rivers, the black flies, Simuliidae, which spread the agents of onchocerciasis, greatly increased the disease burden among local people. River blindness or onchocerciasis has long threatened large swaths of Africa and only recently have been controlled, but has not been eliminated (Budden 1957; Crump 2012). Likewise, the related debilitating disease lymphatic filariasis has long been a scourge of Africa with the earliest known records probably written on Egyptian papyrus (Hajdu 2002). This parasitic disease was spread across the world by Europeans and the slave trade along with *Culex* spp. and other possible vectors. Some vectors such as *Anopheles* and *Mansonia* are not spread onboard ships with the efficiency of some *Culex* spp. or *Aedes aegypti.*

African trypanosomiasis, also known as African sleeping sickness or nagana, is a parasitic vector-borne disease transmitted by tsetse that is almost exclusively endemic to Africa. Ancient Arabic documents report on African trypanosomiasis throughout what is now known as the "Tsetse Belt," but the first medical summary of the clinical symptoms in neurological patients was published at the end of the Age of Enlightenment *c.* 1734 by a British naval surgeon Aktins (Steverding 2008). Trypanosomiasis remains a threat across Sub-Saharan Africa as a zoonosis that infects livestock (Weny et al. 2017). Europeans transported *Trypanosoma evansi*, a parasite derived from African *Trypanosoma brucei* (Lai et al. 2008), from Africa or Asia to the Americas in the 1800s, if not earlier (Desquesnes et al. 2013).

While the true scramble by European nations to divide and colonize Africa did not start until the late 1800s, the activities and discoveries during the Age of Enlightenment paved the way for their colonization. The discovery of a true antimalarial drugs and the eventual discovery of microbial life set the stage for European colonial powers and medicine. The long-term effects of these colonial actions are still in effect today.

ASIA

Asia is the largest continent and insect-transmitted diseases were well documented across Asia from the high arctic to the tropics. Much of eastern Asia and Europe share borders and close economic, religious, military, cultural, and genetic ties. The cities of Asia were just as plague- and typhus-ridden as those of

Europe. Wars between Asian and European powers repeatedly raged across the continent. For example, a series of wars between the Ottoman Empire and the Republic of Venice spanned almost the entire Age of Enlightenment. Diseases such as plague were rampant in these wars and were even weaponized. Flinging dead bodies with the hope of spreading disease and demoralizing the enemy was a medieval siege practice used in both Russia and Europe. During the Siege of Candia, the Venetians planned to use the liquified spleens and bubo fluids of plague victims in a biological warfare attack (Thalassinou et al. 2015).

Russian and Chinese medical practitioners had extensive experience with plague from *Y. pestis* and presumably could identify the disease correctly. Plague also caused untold suffering across much of urbanized Asia during the Age of Enlightenment. For example, a well-documented outbreak of plague occurred in Jaffa following the Napoleonic armies' siege and conquest of the city. The disease was apparently endemic in mainland Asia for thousands of years and thus sporadic outbreaks in places such as China were not uncommon. The third great plague pandemic probably originated in Asia in the 1850s, just after the end of the Age of Enlightenment and spread to India.

Some of the great outbreaks of plague in Asia included an epidemic in Moscow, Russia in 1770–1771. Control measures were largely unpopular and there were riots or civilian unrest that were forcibly suppressed. The outcome of this plague epidemic and the civil unrest caused economic damage to Moscow and thus Russia for years (Alexander 1974). Unfortunately, many of these plague and typhus epidemics caused economic damage that was poorly calculated. The economic losses from lives of peasants and lower classes were often not fully reported.

Louse-borne epidemic typhus played a critical role in the history of Asia. One of the most significant outbreaks of louse-borne epidemic typhus was the epidemic during the Napoleonic invasion of Russia in 1812 (Peterson 1995). This outbreak, as with others during the Napoleonic wars, shaped world history by preventing France from conquering Russia. Millions of soldiers died during the invasion, but since they came in contact with civilians, they spread the infected lice and disease to native Russians. Epidemic louse-borne typhus was a serious threat to Russians throughout much of the Age of Enlightenment (Patterson 1993).

Malaria, caused by *Plasmodium vivax* s.l., possibly evolved in southeast Asia or neighboring Oceania (Escalante et al. 2005). *Plasmodium falciparum*, however, is more likely to have evolved in Africa. Endemic vivax malaria had been a disease in China and probably tropical and temperate Asia throughout most of the region's history. Malaria was also an ancient problem in Middle Eastern Asia and throughout the Indian subcontinent. As with the rest of the world, this disease was particularly dangerous to newcomers, such as European sailors and soldiers in ports or garrisons and to colonists and slaves. Some native people showed resistance to the disease. While India was eventually a major source of cinchona bark for antimalarial treatments, the plantations were not planted until after the Age of Enlightenment in the 1850s.

The colonial exploitation of India was chronically hampered by vector-borne disease and these losses pushed many advancements in medical entomology in subsequent decades. The British-run East India Company founded in 1600 established itself in India and was the de facto ruler of the subcontinent from 1757–1858. Early on in the occupation of India there were purported outbreaks of plague, which were possible, but unsubstantiated due to poor records (Mushtaq 2009). Two diseases, leishmaniasis and malaria, however, were significant problems throughout the history of the subcontinent (Mushtaq 2009). Indeed, the early great discoveries in the ecology of these diseases both came from British officers in India. Ronald Ross, William Leishman, and Charles Donovan discovered the agents or vectors of leishmaniasis and malaria independently while working in India in the late 1800s. The history of India continued to be greatly influenced into modern times by malaria and other vector-borne diseases. These vector-borne diseases were established in the subcontinent and caused significant economic problems for the dominant India in the East India Company.

AUSTRALIA AND OCEANIA

European and Asian explorers did not discover many of the Pacific islands or their native inhabitants until the late eighteenth century. For example, Cook did not land in Hawaii until 1778. Prior to the arrival of the colonial European powers the islands of the Pacific were colonized by small groups of people. This probably provided protection from many of the endemic diseases of the tropics. However, the arrival of Asian and European whalers and naval ships were tied to the extinction of much of the local animal fauna, introduction of exotic animals, and the loss of native cultures. With the destruction of local fauna some endemic arboviruses might have disappeared. Ships also introduced rodents even if they were already present, some feral livestock, and eventually European colonists. Murine typhus almost certainly followed the Europeans to these islands along with rodent reservoirs (*Rattus* and *Mus* spp.), but the long sea voyage required to reach many isolated islands probably protected native people against dengue, yellow fever, plague, and some malaria. The lack of adequate habitat for *Anopheles* spp., however, prevented the establishment of malaria on many Oceanic atolls. While there was an active slave trade between the islands it was not related to the African slave trade and thus did not spread malaria, yellow fever, or similar diseases of African origin as it did in the Americas.

Australia and the isolated Oceanic Islands are perhaps the most enigmatic locals for the early history of medical entomology. The sporadic ancient colonization of the islands was initiated by small groups of people. Anthropologic and archeological studies have documented very little in terms of vector-borne diseases (e.g., Abbie 1960). Zoonotic arboviruses endemic to Oceania and Australia almost certainly caused sporadic disease and death in native peoples. Mosquito-borne arboviruses, such as Ross River virus, Murray Valley virus, and Barmah Forest virus, are endemic across much of Australia (Mackenzie et al. 1994; Harley et al. 2001) and it is almost inconceivable that they did not cause sporadic disease or epizootics prior to their discovery and isolation in the twentieth century. Even in the twenty-first century there are problems with identifying these viruses and their associated diseases (Kizu et al. 2019). Rickettsial pathogens were discovered on isolated Oceanic islands in the twentieth century, but possibly persisted there even prior to human colonization since the introduction of their small animal hosts (Reeves et al. 2012). Indeed, Q-fever was first described form Australia and novel strains of scrub typhus are endemic to the continent and isolated oceanic islands (Derrick 1937; Odorico et al. 1998). Scrub typhus and tick-borne diseases, such as Queensland tick typhus and Q-fever, probably caused occasional epidemics including in the shogunate of Japan. However, unlike the rest of the world there is scant information to show how vector-borne diseases affected natives or their history across Australia and Oceania. There are a few notable exceptions, especially malaria and lymphatic filariasis.

Malaria was considered a particularly important threat to European colonial powers in the Malay Archipelago and Papua New Guinea. *Plasmodium falciparum*, the agent of malignant malaria, was possibly introduced by Europeans, but *P. vivax* was already present (Chhem 2013). Outbreaks of malaria among the endemic people were a documented and significant cause of death and disease. European colonial powers recorded mortality among workers or slaves on their plantations. In addition, two or more malarial agents, *Plasmodium vivax* and *Plasmodium knowlesi*, possibly evolved in this region (Schmidt et al. 1977). *Plasmodium vivax* spread across most of the tropical and temperate regions of the world throughout the Age of Enlightenment. This parasite might even be more adapted to transport by ship than the highly virulent *P. falciparum*, which lacks a persistent liver stage.

The other major vector-borne disease in Oceania was lymphatic filariasis. While the earliest historical records of lymphatic filariasis come from ancient Egypt, the disease also is endemic to southeast Asia and the Pacific (Routh and Bhowmik 2007). This disease is caused by several species of filarial nematodes in the Pacific, and the human infecting *Brugia* spp. probably evolved in southeast Asia and the Pacific region. Endemic transmission and infections remained common in the region until the modern ongoing worldwide eradication program. The islands of Oceania are some of the last endemic regions in the world. Lymphatic filariasis is rarely fatal in itself, but the burden of this disease causes incredible loss of quality of life and productivity. Cases were recorded in ancient art, literature, and oral histories.

The Age of Enlightenment saw the end to the isolation of most of the Pacific Islands and Australia. Almost all of the native peoples were exposed to foreign diseases and the region was exploited for resources. While the early history of the region is poorly documented, the long-term ecological and social changes that followed European and Asian colonization are still playing out with arthropod-borne diseases having a major role in the extinction of local fauna and disease among people.

NORTH AMERICA

Throughout the Age of Enlightenment most of North America was being fully colonized by Europeans, Asians, and African slaves. Native North Americans often suffered historic death rates from exotic pathogens, including several vector-borne diseases, such as yellow fever and malaria. Some of the vector-borne diseases of Europe flourished in the North American colonies and were more significant than they were in the Old World. However, neither plague nor epidemic typhus were great killers of Americans like they were in Europe, Africa, or Asia. Perhaps the lack of large, well-entrenched urban areas protected many of the American colonies.

Louse-borne epidemic typhus was a true, but barely realized threat in North America. Unlike the massive outbreaks of Europe that literally crushed entire armies and depopulated towns, outbreaks in North America were sporadic and limited in scope. There were epidemics in Mexico and South America and possibly small outbreaks among European colonists, but for some still unknown reason epidemic louse-borne typhus was rarely a true threat in the United States or Canada (Humphreys 2006). This is an enigma considering the lousy state of American soldiers in North American wars (Miller 1997). Body lice were a common pest and companion to the residents of North America and there were small outbreaks possibly attributed to typhus during the American Revolutionary War. Unfortunately, for historical epidemiology of vector-borne diseases there were also outbreaks of typhoid that could have been misidentified as epidemic typhus (Smith 1980). The historic lack of typhus during American warfare continued into the American Civil War and remains a conundrum for epidemiologists (Miller 1997). Plague also was largely absent from North America and the first large epidemic was not reported until long after the Age of Enlightenment when it emerged in San Francisco in the early 1900s.

The slave trade introduced several vector-borne diseases from Africa to the Americas. Most of these pathogens were either able to persist in humans for long periods of time or were transmittable while on the slave ships on route from Africa. Parasitic diseases, such as malaria, river blindness, elephantiasis, and even African eye worm (loiasis), were brought to the Americas. Some parasites, such as *Wuchereria bancrofti*, the agent of elephantiasis, were introduced with the slave trade, most likely along with its mosquito vectors, *Culex pipiens* and *Culex quinquefasciatus* (Adler and Wills 2003; Huang et al. 2011). Other parasites, such as *Plasmodium* spp., that cause malaria were established across the continent after adapting to native *Anopheles* spp. as vectors. Not all parasites found competent vectors, ecology, or a host or reservoir population required for establishment. *Loa loa* apparently never was established despite potential vectors (*Chrysops* spp.) and the transport of infected Africans (Orihel and Lowrie 1975). One of the lasting scourges to the Middle and South Americas was the importation and establishment of *Onchocerca volvulus*, the agent of human onchocerciasis during the slave trade (Gustavsen et al. 2011). There were soon endemic areas of Mexico and Central America with established transmission of *O. volvulus*, but no local transmission was reported in the USA or Canada because of a lack of viable vectors.

Yellow fever was one of the major diseases of the Americas. The vector, *Aedes aegypti*, adapted to living on water cisterns on sailing ships and invaded port cities and urban zones (Patterson 1992). The vector almost certainly spread across the Americas long before yellow fever was established. Indeed, the state now known as Florida was once referred to as "Mosquito Country" and successful management of urban and sylvatic mosquitoes has always been crucial in allowing urbanization in this region (Carlson et al. 2019). Yellow fever came to the Americas with the slave trade into the Caribbean (Patterson 1992). It was likely introduced several times before it became endemic in both humans and monkeys. It later adapted to sylvatic canopy mosquitoes and a monkey-mosquito transmission cycle. As with many other diseases of the Old World, this virus killed susceptible native people. There also were some failed but calculated attempts to use it as a biological weapon to kill natives by sending soiled materials. However, unlike smallpox or measles, yellow fever is not spread by fomite, but this was not known until the early 1900s.

Yellow fever was critical in curbing Napoleons ambitions toward the New World. Napoleon probably wanted to conquer North America and garrisoned thousands of troops in French Haiti. The outbreaks of yellow fever in Haiti (then Saint Dominigue) that infected over fifty thousand French troops and killed most of them during the 1790s to 1804 allowed a slave revolt to succeed. The disease was a major factor in political and military history. African slaves and people living on the island were largely immune from past exposure. African slaves engaged in a successful full revolt and Napoleon lost interest in conquering North America. Napoleon then sold the Louisiana Territory to the United States of America and yellow fever allowed the realization of American "manifest destiny" by greatly expanding the country. The same yellow fever outbreak in 1801 also crushed the French army and allowed Haiti to declare independence in 1804, making it the one of first postcolonial independent countries in the New World (Marr and Cathey 2013). As a result, Haiti was one of the first colonies to successfully declare independence form a European colonial power and owes this in large part to yellow fever. This led to complicated relationships between the nascent USA and Haiti as the USA was fearful of slave revolts and also wanted to keep a functional relationship with France.

The endemic zoonotic diseases, such as the arboviral encephalitides, bacterial, and rickettsial pathogens now known across North America probably killed people and caused disease since the initial colonization of North America. However, these are poorly documented and probably focal. Even in the twentieth century, high levels (>10 per cent) of human exposure to some cryptic native arboviral encephalitides were reported from people who did not recall disease (Hammon and Reeves 1952) and in the Age of Enlightenment medical records would have been rare and spotty.

SOUTH AMERICA

European colonial powers claimed most of South America prior to or during the Age of Enlightenment. While the native people of South America suffered massive death rates from European and African diseases, they often managed to fare somewhat better than those in North America. Still, yellow fever, malaria, and ectoparasite-borne rickettsia were a cause of untold deaths. The European colonies exploited and explored most of South America to claim resources, including the first true antimalarial drug. They were also looking for mythological places, such as El Dorado, the city of gold. One colonization of note during the late 1700s was the Falkland Islands, which are almost Antarctic and remarkably never seemed to suffer from louse-borne disease outbreaks, despite the long-term colonization and resulting ecological changes (Armstrong 1994).

As with Europe, the human body louse was endemic across South America. Unlike many human parasites that had possibly gone extinct during the migration of the aboriginal peoples to the Americas, the body louse was present in the Andes region prior to European contact (Raoult et al. 2008). Lice have even been discovered on Pre-Columbian Peruvian mummies (Raoult et al. 2008). The three major louse-borne diseases of humans—epidemic typhus, trench fever, and louse-borne relapsing fever—might have been endemic prior to European contact. There were certainly outbreaks of typhus in the 1600s that killed both European and native peoples across South and Central America. As with North America, there are historical issues with separating typhoid cases from typhus in remote regions of the Americas.

Several vector-borne diseases originated in the tropical regions of the Americas. These diseases have long histories recorded since antiquity in the endemic regions. Many of the diseases, such as Chagas disease, leishmaniasis, and mansonellosis are chronic, endemic, and often cause disease sporadically in susceptible populations. These diseases presented problems along with the introduced Old World diseases for Europeans and their new plantations and mines in South America.

South America played a crucial role in medical history by leading to one of the most significant breakthroughs in the control and prevention of malaria. Malaria was introduced with the African slave trade and exploded into South America in epidemics in the early to mid-1600s and beyond. The first true antimalarial drug was discovered in South America. The apocryphal myth is the discovery of the bark of cinchona trees of the Andes were reportedly used to save the life of Ana de Osorio, the Countess of Chinchón and wife of the Viceroy of Peru, from malaria (Honigsbaum 2003). While there is strong doubt over the validity of the story, the South American cinchona tree does produce quinine, which was the first truly successful antimalarial drug. Cinchona and antimalarial quinine were introduced to Europe and with it the Europeans could exploit tropical regions and combat malaria during warfare and among slaves and workers. Malaria was a great threat and barrier to the exploitation of southeastern Asia, but with the discovery of cinchona the European colonial powers could exploit all tropical regions of the world when given access to the drug.

Wuchereria bancrofti, the agent of elephantiasis in the Americas, was almost certainly introduced from Africa into the Caribbean. This parasite has subsequently caused incredible misery. The African slave trade brought another filarial nematode to the Americas, *Onchocerca volvulus*, the agent of river blindness (Gustavsen et al. 2011). Since slaves that were obviously suffering from parasitic diseases, such as river blindness, were thrown overboard and killed during the transatlantic crossing, only those with less-obvious infections survived the trip and as a result this selected for a unique genetic makeup in the *O. volvus* that invaded the New World (Crump et al. 2012). In addition, the native *Simulium* spp. that are vectors in Africa were not found in the New World and during the invasion of the Americas *O. volvus* had to rapidly adapt to native *Simulium* spp. in the tropics. River blindness in the New World tropics became established in endemic pockets across the continent and caused chronic disease. Once established, this parasite has remained an endemic threat to Native American populations until the twenty-first century when it was possibly eradicated (Boatin and Richards 2006).

CONCLUSION

The Age of Enlightenment saw a rapid spread of European and Asian cultures across the world. The British, Spanish, French, Russian, Dutch, Portuguese, and others spread colonies across much of the globe. By the end of the 1800s, colonies, conquerors, and traders had spread to every inhabited continent and sailed to the coast of Antarctica. Urbanization and population growth across the world along with the spread of slaves and forced labor across the tropical and subtropical regions of the world further spread and enhanced epidemics. Arthropod-borne diseases spread with urbanization, especially in regards to louse-, flea-, and urban-mosquito-borne diseases. Malaria and yellow fever were causes of great death and suffering in both native and colonial people. Some of these diseases also allowed colonies to successfully rebel and declare independence from their European masters, such as in Haiti. Diseases such as plague and epidemic louse-borne typhus killed more Europeans and West Asians in prolonged wars than bullets, bombs, and swords. They prevented the expansion of Napoleon's grand army into Russia and the Americas. Massive civilian deaths in crowded European cities probably played roles in encouraging colonization of places such as North America. These diseases and several others spread across the Americas, Australia, and Oceanic Islands killing most of the endemic people.

While these diseases were causes of great suffering, the Age of Enlightenment saw true medical breakthroughs, which led to the eventual discovery and control of almost all of these diseases in later centuries. The discovery of microorganisms with microscopes began during the Age of Enlightenment. This led to the eventual discovery of bacteria and viruses in future centuries. Real antiparasitic drugs were discovered and their use began to curb the threat of malaria. Advancements in drugs and understanding of human anatomy instead of magical or hoax cures spread from this age in future centuries. True scientific methods and observational techniques were largely developed in the late seventeenth century and while not universally adopted, even in the modern age, they greatly helped with the discovery and elimination of disease. Without the advancements of the Age of Enlightenment many of the modern medical advancements simply would not exist. However, all of these advancements came at a great price in terms of human misery, death, and disease.

CHAPTER THREE

Insects and Food

FRANK N. EGERTON

The beneficial and detrimental impacts that insects have on food for humans and livestock is arguably one of the most important practical aspects of entomology. For example, locusts (*Schistocerca gregaria*) invading crops in great numbers and leaving virtually nothing behind was known in the Middle East since antiquity (Baron 1972). In China, records of locust attacks (*Locusta migratoria manilensis*) were so precise that a graph of their abundance was published for a thousand years, 957–1956 (Konishi and Ito 1973). However, understanding actions of other insect species that had a less dramatic impact upon crops developed more slowly. This chapter presents an overview of the historic publications and discoveries made during the Age of Enlightenment (1600–1820) that pertain to insects and food.

THE EARLY ENTOMOLOGISTS

The first book devoted entirely to insects, *De Animalibus Insectis Libri VII* (1602), was written by Italian professor Ulisse Aldrovandi (1522–1605). He made and reported his own observations, but also drew upon unreliable accounts from ancient, medieval, and the modern authors of those times (Castellani 1970; Findlen 2000a; d'Aguilar 2006; Egerton 2012). The book is well illustrated and contains the earliest known dichotomic key to distinguish between different kinds of species (Beier 1973). Two histories quoted the key to the different kinds of "insects," which included crustacea and worms (Bodenheimer 1959). Even though his book had little to say about the methods to combat agricultural insect pests, it was still important in advancing an understanding of insects, since the more that was known, the more likely such knowledge could be applied to eventually control pests.

The historic encyclopedia, *Insectorum sive Minimorum Animalium Theatrum*, appeared in London in 1634 and has a complex history. Conrad Gessner (1515–1565) was a Swiss encyclopedist who published *Historia animalium* (5 vols., 1551–1587, 3,515 pp.), but these volumes did not include insects (Wellisch 1984). However, he was working on the sixth volume that was dedicated to insects when he died. The volume then went to a friend, Thomas Penny, who had previously worked on it. Penny acquired notes on insects from another friend, Edward Wotton, and then began a synthesis of their works, but died in 1589 before the encyclopedia was finished. That manuscript went to another friend, Thomas Moffett (1553–1604), who added more notes on insects before he also died (Simpkins 1974; Salmon 2000; d'Aguilar 2006). That manuscript was sold to Sir Theodore Mayerne, who had it published under Moffett's name. It was translated into English in 1658. London author George Ordish devoted a dozen pages in his book *The Constant Pest: A Short History of Pests and their Control* to Moffett's descriptions and remedies (Ordish 1976). For example, victims of locust plagues had realized that although no vegetation was left, locusts did leave eggs, which Moffett recommended plowing under. His methodology likely worked for the eggs that were plowed, but this would have been a modest fraction of total number of eggs that were laid. William Harvey (1578–1657) published his claim, *Omne animal ex ovo*, in 1651 (Harvey 1651); therefore, Moffett's recommendations for locust control to maintain crop health is an early application of understanding insect metamorphosis.

In Europe, microscopes stimulated research at the micro-scale and the field of entomology became more rigorous. The first compound microscopes appeared after 1610 (Daumas 1964). One of the earliest known makers of microscopes was none other than Galileo in 1612. The study of insects with microscopes began with the Italian aristocrat, Federico Cesi (1585–1630) (Drake 1971), and his colleague, Francesco Stelluti (1577–1652) (Keller 1976), who had joined Cesi in founding one of the earliest scientific societies, Accademia dei Lincei (1603). In 1625, they used a microscope (perhaps made by Galileo, another member of the society) to study bees, the subject of their work in *Apiarium*, which is preserved as a manuscript in the Accademia's library. Historian of biology Charles Singer has published an article on *Apiarium* and reproduced its illustrations of bees and their mouthparts, which are perhaps the earliest drawings of insects from microscopic examinations (Singer 1953).

The first natural history encyclopedia in Japan was written and illustrated by Nakamura-Tekisai (1666) (Konishi and Ito 1973). Kaibara-Atunobu (Ekken) published a natural history of Japan (1708, 16 vols. + 2 vols. appendices + 2 vols. illustrations, 1715), which described 1,362 materials, including 358 insects first described in this work. "Although he knew that mosquitoes emerged from wigglers, and butterflies and moths from caterpillars, he did not observe that these larvae were born from eggs" (Konishi and Ito 1973). In his *Textbook of Natural History* (1803–1806), Ono-Motohiro (Ranzan) reported that fireflies develop from eggs, "through aquatic larva to adult" (Konishi and Ito 1973). Japanese entomology continued to develop from these beginnings and can be followed in M. Konishi and Y. Ito, *Early Entomology in East Asia* (1973).

Englishman Robert Hooke (1635–1703) led the way in advancing microscopy in his famous published research, *Micrographia, or Some Physiological Descriptions of Minute Bodies Made by Magnifying Glasses* (1665) (Figure 3.1). He lived in London and had been Robert Boyle's laboratory assistant

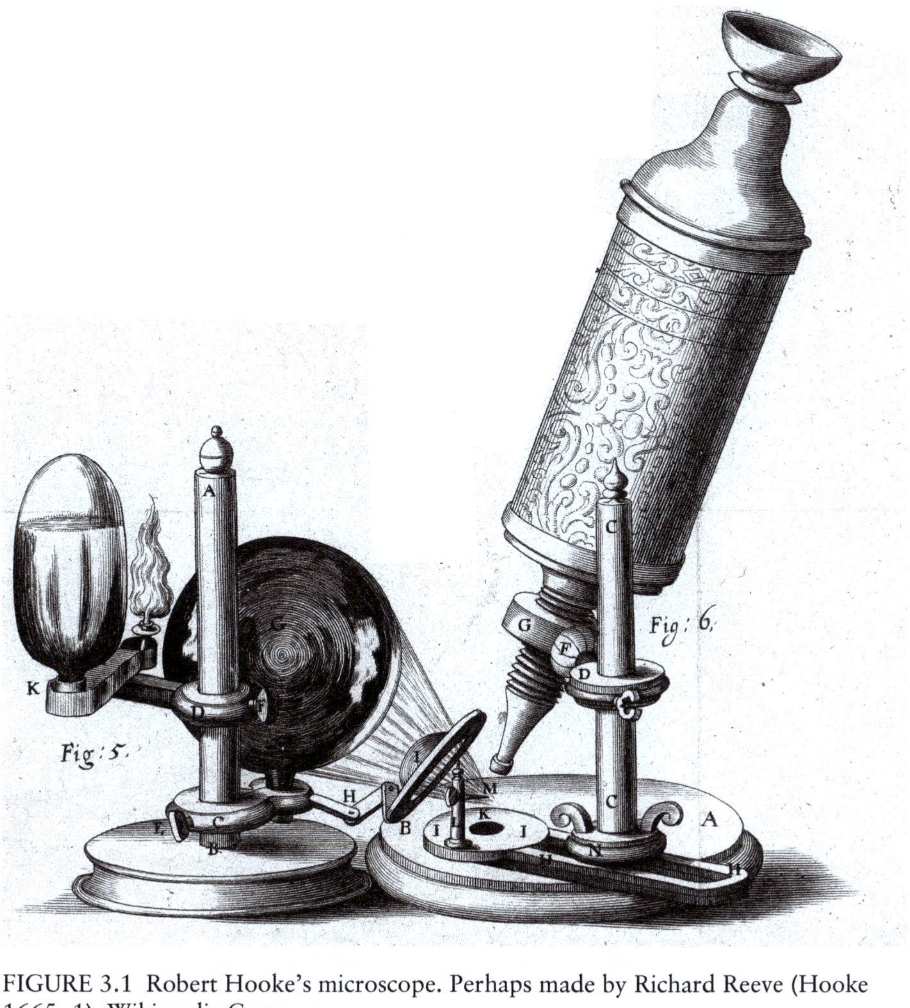

FIGURE 3.1 Robert Hooke's microscope. Perhaps made by Richard Reeve (Hooke 1665: 1). Wikimedia Commons.

at Oxford University (Jardine 2003; Egerton 2005a, 2012; Cooper and Hunter 2006). He devoted a substantial portion of his book to insects (pp. 170–217), but none of the species he discussed and illustrated were crop pests. He did, however, discuss and illustrate familiar parasites of humans: mosquitoes, fleas, and lice (Hooke 1665). He commented on the blood sucking of fleas and lice, but overlooked this in mosquitoes. He studied larval mosquitoes (Figure 3.2), and he commented on their metamorphosis into insects that flew away, but his curiosity did not extend to the adult behavior. Hooke's *Micrographia* was the

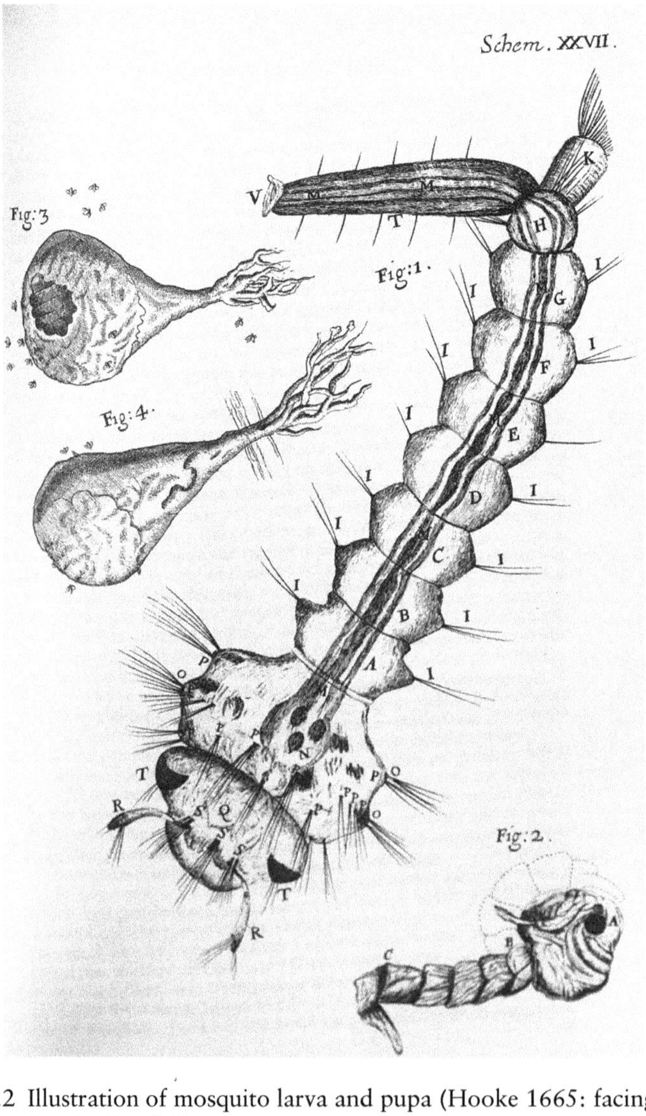

FIGURE 3.2 Illustration of mosquito larva and pupa (Hooke 1665: facing p. 186). Photograph taken by Frank Egerton.

first book on microscopic studies and a convincing display of the power of the microscope in revealing the important secrets of nature, and other naturalists soon followed his lead. Hooke had worked within the context of the Royal Society of London for Improving Natural Knowledge (founded 1660), and that detail also resonated.

An important component of the Age of Enlightenment was that science became rigorous, particularly in its methodology, which was facilitated by Francesco Redi (1626–1698) who invented the controlled experiment (Beier 1973; Belloni 1975; Findlen 2000b, 2000c; Egerton 2005b; d'Aguilar 2006; Egerton 2012). Redi was influenced by both Harvey's and Hooke's books, and he was also a founding member of the Accademia del Cimento in 1657, cimento meaning experiment, in Florence. He was attracted by a controversy over whether some insects gave birth to maggots or only to eggs, which led to his famous experiment. In one vase he placed uncovered meat and in another vase he placed meat covered by a fine net (Redi 1668, 1909). Maggots appeared on the uncovered meat but not on the covered meat. However, Redi was unable to solve the mystery of plant galls, which could have no hole, yet have an insect larva at its center. That mystery would be solved later by fellow Italian investigator Marcello Malpighi (1628–1694). Malpighi was a physician, whose interests included animal and plant anatomy (Adelmann 1966; Beier 1973; Belloni 1974; Malpighi and Adelmann 1975; Meli 1997, 2000; Egerton 2005b; d'Aguilar 2006; Egerton 2012). He discredited the idea of spontaneous generation of insects by asking why they would have sex organs if that were so. He also argued that an insect may lay an egg on a plant and then stimulate it to grow into the gall. He was attacked for this reasoning, and he replied at length in his *Opera postuma* (1697).

The Netherlands was very prosperous during the Age of Enlightenment, and it was during this time that two of the most important naturalists made their lasting contributions to the understanding of insects, Jan Swammerdam (1637–1680) and Antoni van Leeuwenhoek (1632–1723). Although Leeuwenhoek was older, Swammerdam began publishing first.

Swammerdam's father was a wealthy Amsterdam apothecary who had a collection of "rarities" from the East and West Indies. Jan followed in his father's footsteps and made his own collection of insects, which he took to the University of Leiden in 1661, where it impressed fellow medical students and faculty (Schierbeek 1967; Lindeboom 1975; Winsor 1976; Fournier 1990; d'Aguilar 2006; Egerton 2012). Although he earned his MD in 1667, he practiced zoology, not medicine. He was the first to study morphological changes in insect lives, and he classified species according to their life cycles. He made his own instruments for dissections and drew finer details in his illustrations than anyone else at that time. Swammerdam's *Historia insectorum generalis, ofte, algemeene verhandeling von bloedeloose dierkens* (General

Account of Bloodless Animalculae) in 1669 (summarized by Schierbeek 1967) described the development of different species, argued that a pupa was not an egg, but a later stage of development, that plant galls result from eggs laid by insects, and parasitic "worms" found in caterpillars were offspring of other insects. He died from malaria just after his forty-third birthday and left manuscripts eventually published in Dutch and Latin on facing pages, entitled *Biblia Naturae, sive Historia Insectorum* (3 vols., 1737–1738). His dissections of honey bees revealed that the bee "king" was actually a queen (Figure 3.3). *Biblia Naturae* is "one of the outstanding classics of the literature of zoology" (Cole 1944).

John Ray (1623–1705) had broad interests in natural history and he was the most influential English naturalist of his time (Raven 1942; Baldwin 1986; Mandelbrote 2004; Egerton 2005c; d'Aguilar 2006; Egerton 2012). He attended Cambridge University and while there wrote a county flora, the first one in England (1660). His main commitment was to plants, but he was interested

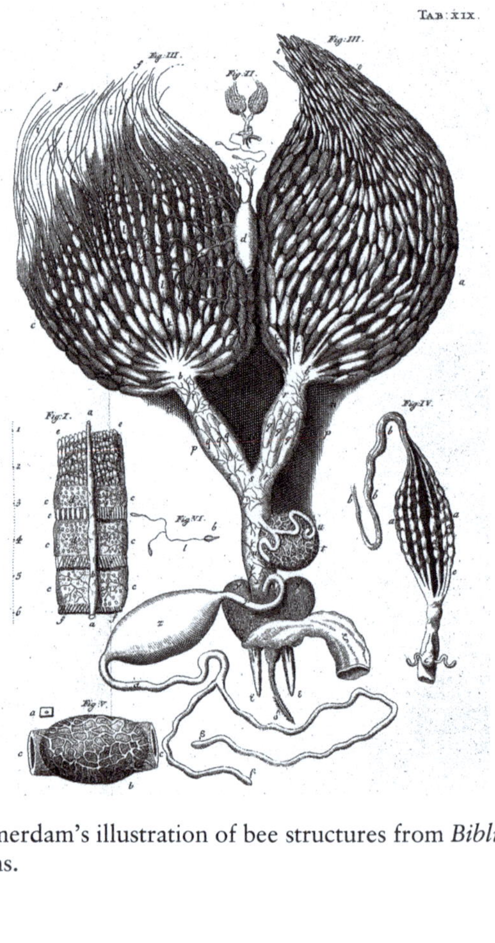

FIGURE 3.3 Swammerdam's illustration of bee structures from *Biblia Naturae*. Wikimedia Commons.

in the interactions between plants and humans or animals. For example, he stated that snails and slugs eat deadly nightshade (*Atropa belladonna*) despite its poison (Ray 1660, 1975). He also reported on other insects that feed on specific plants; the plant species were identified, but caterpillars were of unknown species. Some of the observations in 1660 were discussed in more detail in his *Historia Insectorum* (1710) (Egerton 2005c). For example, in 1660, he could not account for the fact that flies sometimes came out of butterfly chrysalises. However, his *Historia Insectorum* had the answer: "I think that the ichneumon wasps prick these caterpillars with the hollow tube of their ovipositor and insert eggs into their bodies" (Ray 1710, trans. by Raven 1942).

Antonie van Leeuwenhoek (1632–1721) (Figure 3.4) was son of a basket maker in Delft, Netherlands who died when his son was six (Dobell 1932; Amsterdam and Schierbeek 1959; Heniger 1973; Applebaum 2000; Palm 2000; d'Aguilar 2006; Egerton 2006a, 2012). In 1648, he was apprenticed to a cloth merchant in Amsterdam. He returned home about 1654, married, and opened

FIGURE 3.4 Portrait of Antonie van Leeuwenhoek by Jan Verkolje. Wikimedia Commons.

a shop to sell cloth, thread, buttons, and more. Leeuwenhoek attended the dissections performed by a neighbor, who was a physician and city anatomist. Later, he saw a copy of Hooke's *Micrographia* and was intrigued with its illustrations. In 1673 he began making lenses and microscopes (Figure 3.5). A physician wrote to the Royal Society of London and claimed Leeuwenhoek made microscopes superior to all others. He enclosed Leeuwenhoek's account of five observations with drawings. The secretary of the Royal Society was able to translate the letter from Dutch to English, which was then published in the Society's *Philosophical Transactions* (Leeuwenhoek 1700). Leeuwenhoek made unique single-lens microscopes that were more powerful than contemporary

FIGURE 3.5 A replica of one of Leeuwenhoek's microscopes. This image is attributed to Jeroen Rouwkema and is licensed under the Creative Commons Attribution-Share Alike 3.0 Unported license, https://commons.wikimedia.org/wiki/File:Leeuwenhoek_Microscope.png.

double-lens models, with which he studied numerous insects and discovered bacteria, protozoa, spermatozoa, and eggs, both fertilized and unfertilized.

The German botanist Rudolph Camerarius (1665–1721) was son of a professor of medicine at Tübingen, and he also studied medicine and succeeded to his father's professorship when he died (Mägdefrau 1978). He made one important discovery based upon experimentation—that plants have sexual reproduction, and that pistils contain seeds and stamen contain male pollen. He discovered that some species, such as mulberry, have female and male organs on different plants. He assumed that when the female and male organs occur in the same flower that there was self-fertilization, which was later found to be incorrect. He published his discoveries in *Epistola ... de sexu plantarum* (Tübingen 1694, English translation in Bodenheimer 1959).

An interest in insect species and their corresponding food plants also was of interest to artists, which fascinated them, and because their interest began after scientific interest, it seems likely that the naturalists influenced the artists. Arguably the best of these artists was Maria Sibylla Merian (1647–1717) (Figure 3.6), daughter of a prominent Frankfurt publisher of

FIGURE 3.6 Maria Sibylla Merian from about 1700, based on a copperplate by Jacobus Houbraken from a portrait by Georg Gsell. Source is a scan by staedelmuseum.de. Wikimedia Commons.

well-illustrated books (Beier 1973; d'Aguilar 2006; Todd 2007; Egerton 2008, 2012). At thirteen, she raised and painted silkworms, and in 1679, she published *Der Raupen wunderbare Verwandelung* (The Caterpillar's Wonderous Metamorphosis), with a second volume in 1683. She illustrated eggs, caterpillars, cocoon, and adults, which had not been customary, and although she did not use scientific names, her drawings were precise enough for others to supply them. Importantly, she delineated the plants on which caterpillars fed with equal precision; therefore, representing pioneering work about insects and their respective host plants. She and her daughters later lived in Paramaribo, on the Surinam River, where she collected plants and insects, took notes, and made sketches for a new book, *Metamorphosis Insectorum Surinamensium* in 1705 (Figure 3.7).

Antonio Vallisnieri (1661–1730) was son of an Italian official, and had an early interest in nature, which led him to study under Malpighi (Beier 1973; Montalenti 1976; Rappaport 1991; d'Aguilar 2006; Egerton 2008, 2012). He also came under the influence of Galileo. He published an article on insect reproduction in 1700 that was impressive enough for him to be

FIGURE 3.7 Hand-colored copper engraving from Merian's book *Metamorphosis Insectorum Surinamensium*, *Anacardium occidentale*, Plate XVI, 1705. Wikimedia Commons.

appointed to the medical faculty of the University of Padua. In 1713 he published a book on it, which also contained information about four species of flies that parasitize rosebushes (Vallisneri 1713, trans. by Bigelow and Bigelow 1932). He discussed in most detail the garden-rose sawfly (now, *Arge pagana*). On May 6 he observed one land on a young shoot and insert a "hooked sting" (ovipositor) to saw a slit into which she laid her eggs. He cut off the shoot and took it inside, and on May 20 three larvae hatched, and the others emerged the following day. They ate, grew, and moulted several times before spinning cocoons in the bottom of the box. He then found others in the soil below the garden-rose. He described in detail the fly's natural history, including their predators. He ended by discrediting Redi's idea of insects arising from inside plants (though others had done that earlier, and Redi had accepted the new idea).

The versatile René-Antoine Réaumur (1683–1756) (Figure 3.8) was the leading French scientist of his time and the foremost investigator of insects

FIGURE 3.8 Rene-Antoine Réaumur, line engraving by Metzeroth. Wikimedia Commons.

during the 1700s (Torlais 1961; Tuxen 1973; d'Aguilar 2006; Egerton 2006b, 2012). He contributed to mathematics, physics, and technology, but was uninterested in systematics—naming and classifying insects. Instead, he fixed his attention upon species that were significant for human welfare. His first study was on the clothes moth. His *Mémoires pour server à l'histoire des insects* (6 vols., 1734–1742) is an encyclopedia; for example, volume 1 contained fourteen memoirs, 654 pages, and thirty plates. In his second volume (1736), he estimated population increase for a familiar butterfly. He thought it reproduced seven times a year and laid an average of ten eggs per generation. If all survived, he calculated there should be 195,310 per year from this one line. Since no such numbers were produced, he concluded that parasitic worms ate immature forms, which may be part of the answer in some cases. He began work on a seventh volume, and two memoirs, but he died before they were completed (Réaumur 1926, 1955).

ENTOMOLOGY AND ITS APPLICATIONS TO CULTURE

What we can see, in retrospect, is progress in understanding insects from Aldrovandi to Réaumur. Was there progress also in understanding of human food in relation to insects? Moffett's suggestion of plowing under locust eggs was a valid practice, but it would not have destroyed enough eggs to make much of a difference. There is little known evidence of other remedies in the published literature. Microscopy was a very promising subject that Hooke, Swammerdam, and Leeuwenhoek all advanced in different ways. Camerarius's demonstration of sex in plants was not with a microscope, and it did not yet have relevance for insects. Still, the momentum of increasing knowledge on insect biology in the 1600s continued unabated during the 1700s and enlisted the efforts of many more naturalists than before.

One indication of an increased number of naturalists was a corresponding rise in the number of societies to promote such knowledge. During the 1600s there had been important progress in establishing national scientific societies, such as the Royal Society of London. During the 1700s, those general national scientific societies continued to flourish, but in addition, other societies more limited in scope also appeared. English entomologist George Ordish and amateur lepidopterist Michael Salmon have discussed the first entomological society in the world, the Aurelian Society in London, which Salmon traced to about 1738 (Ordish 1976; Salmon 2000). In 1748, a fire broke out while a meeting was in progress and destroyed the societies' records and collections. Another Aurelian Society arose in 1762, but did not last. In 1780, an Entomological Society of London arose and lasted until 1822. Finally, the existing Royal Entomological Society formed in 1833. This faltering history was likely repeated in other countries.

The earliest universities arose during the Middle Ages, and their numbers have increased steadily ever since. During the Age of Enlightenment, a university would most likely have a medical education program, with one of its professors having some biological interests. Such a university might also have a professor of natural history. One outstanding example was at the University of Uppsala, which employed Professor Carl Linnaeus (1707–1778) (Blunt 1971; Tuxen 1973; Frängsmyr 1983; Kroner 1999; d'Aguilar 2006; Egerton 2007, 2012) (Figure 3.9). Although Sweden had a much smaller population than France,

FIGURE 3.9 Carl Linnaeus, painted by Alexander Roslin. Wikimedia Commons.

Linnaeus achieved even more renown than Réaumur had. His publications were more numerous than Réaumur's, and their total pages published was probably similar (Soulsby 1933). Linnaeus attended the University of Uppsala and obtained a botanical education, but he went to the Netherlands in 1735 to earn his MD. For him, that degree was little more than a qualification for teaching in a university. He stayed in the Netherlands for three years. In 1735, he published his most famous work, *Systema Naturae*, in which he regularized giving every species a generic, then specific name, such as *Homo sapiens*. He also needed to classify related species, and for flowering plants he devised a key based upon the number of pistils and stamen in flowers (qtd. in Hall 1951; Bodenheimer 1959). Linnaeus's system enabled entomologists to achieve order in their attempt to identify the legions of insects already discovered, and his significant impact on all biological fields was why he was designated as the lectotype for *Homo sapiens* in 1959.

Slightly after Kaibara-Atunobu's *Natural History of Japan* (1708–1715), other national insect books appeared in Europe (Beier 1973; d'Aguilar 2006). A German teacher, Johann Leonhard Frisch (1666–1743), published *Beschreibung von allerley Inseckten in Teutschland* in thirteen parts (Berlin, 1720–1738), with "an abundance of excellent observations" of about three hundred species, including insect development. He emphasized the study of pest species in order to determine how to combat them. Londoner Eleazar Albin (active 1690–1742) earned his living as an artist and drew the illustrations for his *Natural History of English Insects* (1720) and *Natural History of Spiders and other Curious Insects* (1736). In the latter work he included a portrait of himself on horseback (Salmon 2000). The first edition of his *English Insects* was so expensive he sought wealthy patrons for it. One could buy it with a hundred colored illustrations for three guineas or uncolored for thirty shillings. Frisch and Albin showed that dedicated amateurs also can make useful contributions to entomology.

Swedish Charles de Geer (1720–1778) (Figure 3.10) was born with a silver spoon in his mouth (Bryk 1952; Landin 1972; Tuxen 1973; Egerton 2008, 2012). He grew up in the Netherlands, but at age nineteen he returned to Sweden to claim his inheritance. He was elected to the Swedish Academy of Sciences, for he was very wealthy. He began publishing articles on insects in the Academy's transactions in 1740, first on Collembola. He took the time to manage his estates and become the wealthiest Swede, but his heroes were Leeuwenhoek, Swammerdam, and Réaumur. He took Réaumur's title as his own and he drew the insects illustrated. Réaumur persuaded him to publish in French his *Mémoires pour servir à l'histoire des insects* (7 vols., 1752–1778, German 1776–1783). His interest was in insect behavior, and not in naming and describing. His insect collections are now in the Swedish Museum of Natural History.

FIGURE 3.10 Charles de Geer, painted by Gustaf Lundberg. Wikimedia Commons.

From the Geneva area (not then part of Switzerland), Charles Bonnet (1720–1793) (Figure 3.11) was born in the same year as de Geer, and like de Geer, Bonnet corresponded with Réaumur (Savioz 1948; Pilet 1970; Tuxen 1973; d'Aguilar 2006; Egerton 2008, 2012). Bonnet was handicapped by

FIGURE 3.11 Portrait of Charles Bonnet in 1777, by Johan Frederik Clemens. Wikimedia Commons.

physical disabilities; as a child, his deafness isolated him from his peers, and by about age twenty-five he also began losing his sight. However, before that, Réaumur proposed an experiment to him: could he raise a plant louse (aphid) alone and see if it would reproduce without mating. Bonnet isolated an aphid under a glass vase with a little branch on May 20, and it shed its skin four times by May 31 (Hall 1951; Bodenheimer 1959). On June 1 it began giving birth to live young, and by June 21 it gave birth to ninety-five young. He had discovered parthenogenesis. He published his *Traité d'Insectologie* in two volumes in 1745, while he was going blind.

Denmark had two notable entomologists by the mid-1700s. Otto Frederik Müller (1730–1784) was from Copenhagen and attended its university before becoming a tutor for the son of a nobleman (Spärck 1932; Snorrason 1974; d'Aguilar 2006; Egerton 2008, 2012). He traveled with the family around Europe, became interested in natural history, and published *Fauna Insectorum Fridrichsdalina* (1764), which described 858 insects, spiders, and centipedes on the estate where he tutored. He followed that with *Flora Fridrichsdaling* (1766) which described 1,100 species. In 1773 he married into wealth and devoted his time to studying plants and animals, culminating in *Zoologiae Danicae* (2 vols., 1779–1784). He became known as the Danish Linnaeus.

That recognition fit Johann Christian Fabricius (1745–1808) even better, since he studied under Linnaeus for two years (Landin 1971; d'Aguilar 2006; Egerton 2008, 2012). Although Danish, in 1775 he became professor of natural history at the University of Kiel, now in Germany, where he remained. His interest in insects was worldwide, and every year he studied specimens and literature at other west European universities and museums (Armitage 1958). His vast publications, in German or Latin, were descriptive and systematic. He named and described some ten thousand insect species (Essig 1931; Tuxen 1973). His *Systema Entomologicae* (1775) replaced Linnaeus's classification. Soren Tuxen judged Fabricius's *Philosophia Entomologica* (1778), patterned after Linnaeus's *Philosophia Botanica* (1751), his "most important book of all and one of the most important books in entomology of all times" (Tuxen 1967).

If Fabricius is considered the leading entomologist of his generation, that honor belonged to the Frenchman Pierre-André Latreille (1762–1833) in the following generation (Burkhardt Jr. 1973; Lindroth 1973; d'Aguilar 2006). Working in Paris, Latreille had the advantages of more resources and colleagues than Fabricius had in Keil, and he constructed a better classification than before. However, for this chapter, I will not focus on the developments of systematic entomology.

Insect behavior became more relevant for relations to human food. Joseph Koelreuter (1733–1806) was son of an apothecary who graduated from the University of Tübingen with a medical degree in 1755 and then became keeper of the natural history collections of the Imperial Academy of Sciences in St. Petersburg (Olby 1966, 1973; Brown 1973). He began to study flowers and pollination, and the Academy offered a prize in 1759 for an essay explaining an experimental demonstration of the sexuality of plants. That focused his research, but he did not finish his study in time to win the prize, which went to Linnaeus for two reported experiments. Koelreuter returned to Germany in 1761 and began publishing his findings in Leipzig, as he did again in 1763, 1764, and 1766. His emphasis was on hybridization, and he initiated experimental genetics. His studies of pollination were in support of his experiments.

Christian Konrad Sprengel (1750–1816) was a fellow German who became head of a Lutheran school in Spandau (Brown 1973; King 1975; d'Aguilar 2006). In the summer of 1787, he began observing pollination of geranium flowers, and he continued pollination studies for six years, leading to the publication in 1793 of his *Das entdeckte Geheimniss der Natur im Bau und in der Befruchtung der Blumen* (The Discovered Secret of Nature in the Structure and Fertilization of Flowers) (Bodenheimer 1959). It contained 1,117 drawings of floral parts of 461 species. His secret of nature was that flowers are shaped to attract specific insects to their nectar, and they take pollen from one flower to another to mutual advantage.

Sprengel's exact contemporary, François Huber (1750–1831), was from Geneva, and his father, Jean Huber, wrote *Obervations sur le Vol des Oiseaux* (Geneva 1784). When François was fifteen, he began going blind, but with the aid of his wife and servant, he was able to study the honey bee (Richard 1973). His instruction was from Réaumur's *Mémoires*, and he also corresponded with Bonnet, who suggested experiments on bees and making beehives from glass (he improved on Réaumur's specifications). His *Nouvelles Observations sur les Abeilles* (Geneva 1792) described the life history of honey bees in greater detail than anyone previously had, and in 1814 he published a second volume which discussed various subjects, such as the origin of wax, comb construction, respiration experiment, and senses. Volume one was translated into English in 1806 and is available online. His son, Jean Pierre Huber, followed in his father's footsteps, but chose to study ants.

Local and national accounts of insects continued to be published; undoubtedly the most detailed was Edward Donovan's *The Natural History of British Insects* (16 vols., 1792–1813, 569 color and seven black and white plates). Yet, he was not an academic professional (Salmon 2000; Gilbert 2004; Wilson 2004; d'Aguilar 2006). He is presumed to have been wealthy, since he devoted so much time publishing illustrated books. Incredibly, he simultaneously also published *General Illustration of Entomology* (3 vols., 1798–1805), *The Natural History of British Birds* (10 vols., 1794–1819), *The Natural History of British Shells* (5 vols., 1799–1803), *The Natural History of British Fishes* (5 vols., 1802–1808), a travel book, and more. Some of his insects are now in the Natural History Museum, London, and in Hope Collections, Oxford.

There were about 250 British entomologists in 1800 (Clark 2009), and most were amateurs. Commercial establishments supplied them with equipment and supplies (Chalmers-Hunt 1994). One amateur naturalist, though not an entomologist, was Erasmus Darwin (1731–1802), prominent physician and intellectual (King-Hele 1963; Cohen of Birkenhead 1971; Smith and Arnott 2005). One of his books was *Phytologia, or the Philosophy of Agriculture and*

Gardening (1800), which included pests and possible remedies (pp. 315–72). Ordish (1976) evaluated Darwin's discussion, summarized here: aphids reproduced either by eggs or live young, and washing away eggs with water was an inadequate remedy; experimentation was needed, and he suggested artificial propagation of its predators, lacewing flies and ladybirds. To preserve cabbages, use nets to catch and kill white butterflies. In France, corn butterflies (grain moth, *Sitotroga cerealella*) were controlled by heating. He thought bees and moths were injurious to flowers, because robbing nectaries of honey deprived anthers and stigmas of nourishment. In the seven years since Sprengel had published *Geheimniss der Natur* (1793), word of his discoveries had clearly not reached Darwin. Water rats caused much damage; they had two entrances to their nests, one below water and the other in tall grass and weeds on the bank. If one cut plants on the bank, they would move elsewhere. American water rats had tapeworms; if some of those rats were imported, they might infect English water rats.

INTERACTIONS BETWEEN PEOPLE AND INSECTS: THE CONTRIBUTIONS OF KIRBY AND SPENCE

The most prominent British entomologists in the early 1800s were the amateurs William Kirby (1759–1850) (Figure 3.12) and William Spence (1783–1859) (Figure 3.13) (Salmon 2000; Clark 2004a, 2004b, 2004c). Kirby graduated from Cambridge University and became a country parson, which left time for studying insects. Spence was a merchant. They did not let academia organize their joint presentation, but addressed readers who cope with insects, in *An Introduction to Entomology; or, Elements of the Natural History of Insects: comprising an Account of Noxious and Useful Insects* (4 vols., 1815–1826; edn. 6, 2 vols., 1843) (plate shown in Figure 3.14).

They devoted five chapters (called "letters") to injuries of people, beginning with a discussion of insects that attack humans directly, organized under three headings: 1) "seek to make us their food"; 2) "seek to prevent or revenge an injury which they either fear, or have received from us"; 3) "offer us no violence, but yet incommode us extremely in other ways" (Kirby and Spence 1843). They began this account with "a very disgusting genus, which Providence seems to have created to punish inattention to personal cleanliness." They cite Leeuwenhoek who found lice nits (eggs) hatch eight days after being laid and begin laying eggs when a month old. That meant that one female louse (*Pediculus*) would have five thousand descendants in eight weeks. If one has many lice, it causes extreme irritation, called phthiriasis.

Mites (*Acarus*) may cause Ophthalmia, also characterized by severe itching. They cited Linnaeus as believing that "many contagious diseases are caused by

FIGURE 3.12 Lithograph portrait of Reverend William Kirby, President of the Ipswich Museum, by T.H. Maguire, 1851. Wellcome Library no. 5028i. Photo number: V0003225. Wikimedia Commons.

FIGURE 3.13 William Spence, engraving by William Thomas Fry. Wikimedia Commons.

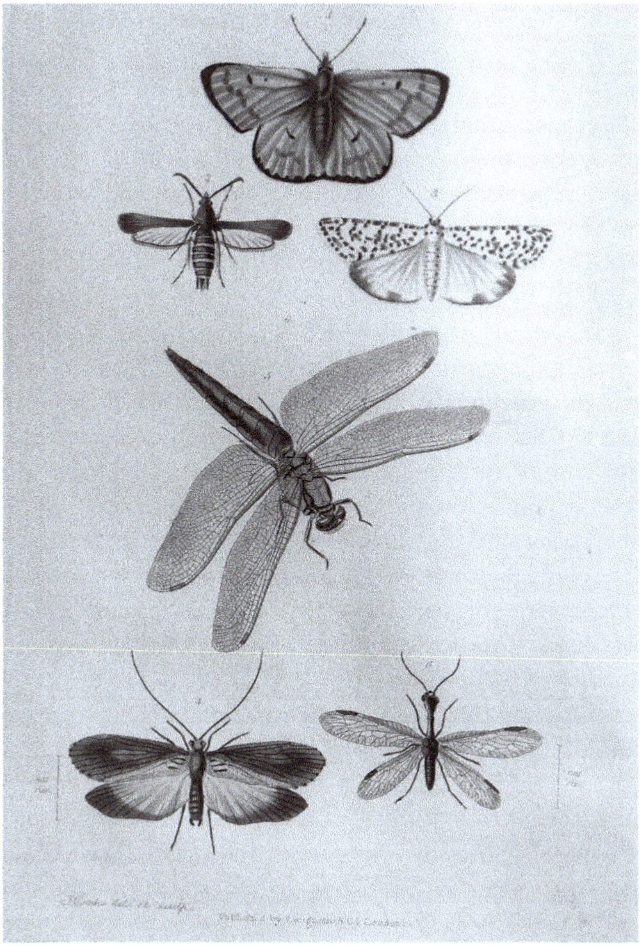

FIGURE 3.14 Plate 3, figure 4: Trichoptera (*Phryganea varia*); figure 5: Neuroptera (*Libellula cancellate*); figure 6: ditto *Raphidia notata* (all from vol. 1). Kirby and Spence's plates of insects. Photograph acquired by Frank Egerton.

mites" (Kirby and Spence 1843). Scabies had been blamed on mites since the Middle Ages; however, perhaps not all cases of scabies are caused by mites (Kirby and Spence 1843). They cited other possible examples, including "pigmy tormentors" that were fleas (*Pulex irritans*). Hungarian shepherds repelled fleas and lice by coating their linen with hog grease (Kirby and Spence 1843). Then discussions of chigoes, ticks, bed bugs, horse flies, and mosquitoes. The last being our greatest pests (Kirby and Spence 1843; Cameron 1973; Lindroth 1973), and that was before it was known that some mosquitoes transmit diseases. Insects that attack from revenge or fear were bees, wasps, ants, scorpions, centipedes, and poisonous spiders. Simply annoying insects included minute flies (*Thrips physapus*) and gnats (*Simulium reptans*) (Kirby and Spence 1843). House flies were worse because they were more numerous. Hairs of caterpillars can cause itching or pain (Kirby and Spence 1843).

Kirby and Spence then explained insect injuries to domesticated animals, thus affecting food production. Horse flies (*Tabanus*) and little flies (*Stomoxys calcitrans*) make horses miserable by sucking their blood at wounds they inflict, and horse bees (*Oestrus equi*) lay eggs on hairs, which horses lick and so introduce them internally (Kirby and Spence 1843). Even more troublesome is a related species, *O. haemorrhoidalis*, which deposits eggs on horse lips. In Hungary there is a worse fly (*Rhagio colubaschensis* Fabricius or *Simulium* Latreille) that attacked cattle in great numbers and could be fatal within a few hours (Kirby and Spence 1843). Sheep were plagued by a gadfly, *O. ovis*, that oviposit eggs in their nostrils, and the maggots inhabit the maxillary and frontal sinuses, then fall to the ground at the pupa stage (Kirby and Spence 1843). Sometimes maggots went astray and ended in the brain. Swine were least bothered by insects, having only a louse that caused itching. Laplanders domesticated reindeer, which were tormented by two species of gadfly—*O. nasalis* and *Tabanus tarandinus* (Kirby and Spence 1843). Domestic dogs suffered from fleas, lice, and ticks. (Nothing on cats.)

Honey bees also had their enemies. Wasps and hornets were predators, and there were numerous parasites (Kirby and Spence 1843). They cite a Pediculus, *Melittophagus*, and mites, without details on how they were damaging. The bee louse (*Braula caeca* Nitsch), the size of a flea, "often infests populous hives so as to greatly annoy the bees by fixing itself upon them (sometimes two, three, or more on a single bee), and making them restless and indisposed to their usual labours." Corsair bees (*Philanthus apivorus*) seize worker bees and carry them off to feed their young. Death's head hawkmoths (*Acherontia atropub*) rob hives of honey. They produce a sound that has an effect similar to the voice of their queen, striking them motionless while they rob them of honey. Two other moths, *Galleria cereana* and *Mellonella*, pass their initial growth in hives, unmolested in tubes made of wax. These larvae are sometimes so numerous, the bees desert the hive.

Kirby and Spence (1843) also provided a lengthy survey of insect damage to crops. They began with wheat; a number of insect larvae are discussed, some of which seem clearly established, and some are of debatable identity. The wheat that survives for harvest then has another hazard in storage from the weevil *Caladra granaria*, one of which can produce six thousand descendants in a year (Kirby and Spence 1843). This leads to a discussion of other insects that attack other kinds of stored grain. Where do these accounts lead? They move on to kitchen-garden plants. In other words, they had no remedies to suggest. However, if one identified both the kind of plant and the kind of insect at hand, one would know what kind of damage to expect, and some day someone might find an effective way to combat the insect. Furthermore, there were indirect benefits from some insects. For example:

> ... insects that attack the roots of the grasses, and, as has been before observed, so materially injure our herbage, the wire-worm, the larvae of *Melolontha vulgaris, Tipula oleracea,* &c., in ordinary seasons only devour so much as is necessary to make room for fresh shoots ...
>
> (Kirby and Spence 1843)

And, "it is not improbable that the *Tabani* often act as useful phlebotomists to our full-fed animals ..." and as Voltaire put it in a satire, "This is the best of all possible worlds." A more realistic idea was that some insects limit the abundance of noxious insects. Further, some insects consume the excrements of other animals and also dead animals (Kirby and Spence 1843). Other insects provide food for other animals. The authors describe several kinds of plants that trap and consume insects (Kirby and Spence 1843). By the time Kirby and Spence published, the importance of insects for pollination of many flowering plants was also known, and so described.

There were also direct human benefits from insects, most famously honey from beehives. These authors reported that some Ukrainian peasants had four hundred to five hundred beehives and made more money from them than from raising grain (Kirby and Spence 1843). In Spain, one parish priest had five thousand hives. All European hive bees were *Apis mellifera*, except in parts of Italy, which had *A. m. ligustica* Spinola. In South America, much honey is collected from *Trigona Amalthea*, which build nests in trees, and in Ceylon, natives collected honey from *Bamburos*. In Egypt, a hive bee was *A. fasciata* Latreille. In India, *A. indica* is cultivated.

Europeans gladly eat crustacea, mainly shrimp and lobster, but not insects. Kirby and Spence viewed this preference as merely a tradition (Kirby and Spence 1843). They stated that insects one might eat are species that feed on plants, as explained by Rev. F.W. Hope, *Entomological Society Transactions*, volume 3, part 2. Many Coleoptera larvae were eaten in different places: "The grub of the palm-weevil (*Cordylia palmarum*), which is the size of the thumb, has been long in request in both the Indies" (Kirby and Spence 1843). Larvae of the larger capricorn, *Cerambyx* L. and *Longicornes* Latr., were considered delicacies in many countries. Grubs of *Prianus damicornis*, the thickness of a man's finger, were eaten in Surinam, West Indies, and in America. Their examples go on and on for a dozen pages. Kirby and Spence did not provide any personal testimonials to strengthen claims from sources they cited on the tastes of such insects.

In the West Indies and in South America, natives used fireflies as a lantern (Kirby and Spence 1843). In China, Chile, and Brazil, ladies used parts of certain insects to decorate their dresses. And, of course, silkworms provided threads sturdy enough to weave "silk" cloth. Still other species could be used to produce dyes. They included details on the species and procedures involved.

CONCLUSIONS

During the period 1600–1820, the science of entomology emerged, mainly in Europe. A few examples are cited from Japan. Even more examples could be cited from North America. To say anything significant about insects and food, one needed to identify the insects under discussion, and have readers securely understand which species are meant. That was a formidable task because species of insects are more numerous than any other group of organisms, and even today are not fully identified. Consequently, naturalists had to believe that this challenging task was worth the trouble. Enough of them did believe it to steadily acquire more and more knowledge, and by 1600, printed books were products of a technology (printing press) taken for granted. Some ground rules also had to be agreed upon. They accumulated slowly and without much fuss. Two examples: scientific societies and Linnaeus's binomial system of species names. National scientific societies developed during the 1600s in Britain and in France—where there were nations—and in some smaller countries. In the 1700s, a few entomological societies also developed in these two countries and elsewhere.

Along the way, some life history information accumulated on insect species of practical interest. Identification of relevant plant species was also needed, and botanists were ahead of entomologists, and so this was not much of a problem. Mosquitoes were identified as the worst pest to torture humans and a few domestic species, but the discovery that mosquitoes transmit diseases was in the future. Developing knowledge of insects as causes of crop damage was slow to develop, and it seems reasonable to assume that Kirby and Spence summarized what was known by 1820. There had been some progress by the 1810s in identifying the problem insects, but the remedies they discussed were not very effective. The story of insects as assets to humanity was better understood. Bee keeping had begun in antiquity, yet naturalists in this period discovered that "king" bees were queens, and they learned much about insects that interfere with bees. They also knew about dyes from insects, though naturalists were not the ones who discovered this.

Entomology in 1820 had momentum, and in 1820 one could reasonably expect knowledge of insects and their relations to humans to continue expanding.

CHAPTER FOUR

Insect Products

GENE KRITSKY

The Enlightenment sparked the expansion of human knowledge, which was due, in part, to the increased availability of mass-produced books. These books were made possible with Gutenberg's invention of the moveable type printing press during the Renaissance. Gutenberg's invention was to promote religion with the printing of Bibles, but for that to work people needed to be literate. This, however, turned out to be a double-edged sword because the growing literate public created a market for books beyond the Bible, and it was this demand for books that lit the fuse starting the Enlightenment (Epstein 2008).

The first printed books involving insect products appeared towards the end of the Renaissance and dealt with apiculture. These early books helped spread interest in beekeeping, and several new beekeeping books were in the works at the start of the Enlightenment. One of the earliest such books was *The Feminine Monarchie* published in 1609 by Reverend Charles Butler.

REV. CHARLES BUTLER'S *FEMININE MONARCHY*

Rev. Butler was born in 1563 in Buckinghamshire, England. He attended Magdalen Hall at Oxford and completed his BA in 1583 and his MA in 1587. He spent the bulk of his professional life as a clergyman at the Laurence-Wotton Church where he served for forty-eight years. He wrote books on English grammar and music, but his lasting contribution was his book on honey bees (Pruett 1963).

The Feminine Monarchie documented that Butler was himself a beekeeper. His interest in beekeeping was to increase his income and to raise a dowry for his daughter. His book was well written and one of the first books to have all the appropriate sections of a professional publication of the seventeenth century, such as a dedication and a letter to the reader. The book's content was divided into chapters that were arranged in a logical sequence. He included margin notes to help the reader quickly find specific content or references. Butler kept his bees in skeps, and his book was, according to Fraser (1958), the best source of information on early skep beekeeping at the time, and it remained in print for two hundred years.

Butler (1704) used both straw and wicker hives, and he gave straightforward descriptions and the benefits of both hives at the beginning of Chapter 3:

> Some use straw hives bound with bramble, others wicker hives made of privet, wicker or hazel, daubed with fine loam tempered with sand or small sifted gravel, to bind it and keep it from flawing or cracking in the hot weather; but though these from long custom have been in use, especially amongst the country people, yet there is an ill conveniency in them, as I shall show you as briefly as may be, with some remedies to retrieve it.
>
> The straw hives when they are old and loaded with their combs, do usually sink on one side, especially if they take wet, and so mash and spoil the combs, killing many of the bees in their fall, and spill the honey, for which cause see that they be well and hard wrought, and strengthen them with a strong cope fitted to the top of the hive.
>
> The wicker hives will decay and lie open if they be not often repaired, so that the wasps and other robbers will make their way at the breaches, to plunder the honey, and annoy the bees; nay, the mice if they see but a flaw will work their way into the hive, by gnawing the twigs that obstruct their passage. However, if either of these hives are subject to wet and moldiness, nay, often to rottenness, which will rot the combs and destroy the bees, but the cold in the winter, and the heat in the summer do sooner pierce the wicker hives, for which cause it is proper to double daub them, and repair them as often as you perceive any cracks or flaws in the loam, and indeed the straw hives are better for small swarms: especially, [*sic*]
>
> When the bees hang around together, they in the best manner defend themselves from the cold, and therefore the nearer the hive comes to the fashion of a globe in the roundness, the warmer the bees will keep themselves; but of necessity the bottom must be broad, for the upright and sure standing

of the hive, and for better taking out the combs, and for the better taking out of the combs, …

A handle at the top of each hive is requisite for two reasons, *viz.* carrying off the hive and staying the hackle; the wicker hive is made straight handle which is at the end of the main staff, but the straw hive a bow handle is most convenient, made of small withies or good bramble. Hives are to may be made of any size between five and 7 gallons, that's so they may be proportioned to the swarm, as it is greater or less …

(Butler 1704)

Butler (1704) also detailed how the interior of the hives should be prepared.

When you have your hives in a readiness, the next thing is to know how to dress them.

First then remove all the standing out straws, twigs, or what may be rough or offensive to the bees when fastened to the hive, making the inside as smooth as may be, for such roughness and hindrances often cause the bees to dislike and forsake the hive, yet if they keep to it they will be much troubled and hindered by them till they remove them themselves if they are able, by scraping and gnawing, which you may hear them doing in the nighttime, by laying your ear to the hive, sometimes for a week together, especially in a straw hive, which yourself may remedy in in a few minutes, with a fine sharp knife, pruning them until no ruggedness be left …

In a wicker hive that the upper ends [or ribs] rest against the middle of the staff, and nether ends against the parts of it between the Wickers, and in the straw hives set the upper ends together in a cop, and the nether ends against the brambles or threads, under the third or fourth roll; as for the cop it is a round piece of wood an inch or two thick whose lower surface is flat with a hole in the middle, half an inch deep for the splits or splinters to rest in, and the upper part is convex around turd or made otherwise fit to the cavity or the top of the hive.

To make your splits, take straight willow or hazel stick, quarter it as if it is big enough, otherwise split it only in halves, then smooth the splits so made, out leave them of a convenient strength also of proper length and the lower ends forked, to rest against the sides of the hive …

(Butler 1704)

Butler's readers successfully followed these detailed instructions to prepare their hives. Butler's directions also differed from those published by Edmund Southerne's (1593) in that Butler included a cop or a circular round piece of wood incorporated in the dome of the straw skeps. Three or four thin splits of wood were inserted into a hole in the cop, bowed and inserted between the third or fourth straw coil at the bottom of the hive. This provided strong support for the comb, but also helped to keep the hive from collapsing inward from the weight of the comb when it is filled with honey.

With the hives prepared, Butler (1704) enlightened his readers on where to place the apiary, a group of hives kept in a single location:

> When you intend to place your bees, choose some convenient plot of ground near your house, that they may be in sight and hearing, because of swarming fighting and other sudden accidents, wherein they need assistance to prevent their flying away, or the mischief which may be timely remedied, if you have a few stalls, your garden of flowers or herbs may serve them, but as they increase they require a green plat fitted for the purpose, safe and surely fenced, not only from all kinds of cattle, but also from the violence of the winds, that when they come laden home, and weary with her flight, they may quietly settle. The north fence should be close and high, that the cold wind from that quarter may not offend them, for in the early spring would be apt to kill them or cast them down. In the setting, place them rather on the south side of your house than on any other and the east and the Northeast, and the north west fences be as high as the north one and keep them as much as may be from the sun by shadows when it grows hot, for too much heat hurts them as well as too much cold ... A house, wall or high pail is best for the northern fence ...

Once the location has been determined, Butler addressed how to protect the hives. He wrote that the hives should not be placed on the ground as mice were apt to find their way into the hives. He recommended placing the hives on stools or a bench not more than "two feet above the ground" with some space between the hives to reduce bee conflict. These hives could be under some kind of roof to keep the rain from falling on the hives or collecting on the stools or on the bench and flow in the hive (Butler 1704).

Further protection could be provided with a hackle, a cover for each hive. Hackles helped to insulate the hive from the heat in summer and the cold in winter. Butler (1704) wrote:

> To make this hackle take four or five large handfuls of wheat or rye straw the longest, which is very well cleaned from the corn, that may not grow by the wet falling on it, or to prevent that all together, having laid the straw even it at length and cut off the chaffy heads where the corn grew, then tie

> the small ends of the straw very fast, and cut the other end even, open it then in the middle, and put it over the hive that they stand up like a pyramid or a Sugarloaf that's so it may shoot off the rain like a thatch of a house, also keep off the cold and heat, ... The hackles must be so fitted to the hives, that they may cover them within an inch of the stool then with a round belt or girth gird the hackle close to the hive, that may not be disordered by the wind, and if there be any crack in the hackle or belt, set a foremost, that standing furthest it may shoot the water the better from the door, and let the belt be somewhat higher than other places, and every now and then takeoff the hackle, to see if no mice, moths, earwigs, or other such enemies of the bees lurk or harbor in it, and if there be warm or windy weather after much wet, let it be off for a time to air the hive, and dry the moisture from it, that so molding this may be prevented.
>
> (Butler 1704)

Butler (1704) was a keen observer of honey bee behavior, and he came very close to making a significant discovery about their communication. Butler noted that the swarms will first settle down on a branch. If by chance the queen did not swarm with the bees, the bees would return to the hive. If she was in the swarm, then bees would send out "spies," what we now call scout bees, to find a suitable site for a new hive. When these spies return, "you may perceive it by the shivering and fluttering their wings as if they were chilly, and the sound or the watchword goes from one to another through the whole swarm." Butler noted if you had already hived the swarm when this shivering behavior occurs, then the bees would leave your hive and proceed to a new location. Butler was describing the waggle dance that scout bees perform when they are instructing the swarm of a suitable new cavity within which they should go to take residence. It would be another three centuries until Karl von Frisch would work out the details of what Butler was describing (Brown 1994).

DESIGNING NEW HIVES

Butler kept his bees in both wicker and straw skeps. The increase in beekeeping books early in the Enlightenment, however, documented experimentation with new hive designs, and one of the most innovative in the mid-seventeenth century were octagonal hives. Octagonal hives were made of narrow boards of wood that, when assembled in an octagonal arrangement, created a cylindrical hive. This shape was thought to be more natural for honey bees, because honey bees are cavity nesters, and the most common cavities to be found in the wild were found in hollow trees. Octagonal hives made of boards would seemingly be more natural for the bees.

One of the first of such hives was described in a letter from Christopher Wren, the architect of St. Paul's Cathedral in London, in a letter that was published by Samuel Hartlib in 1655 in his book, *The Reformed Commonwealth of Bees*. This book consisted of a series of letters to Hartlib, which he published along with information on rearing silkworms. Hartlib was originally from Poland and moved to England in 1623, but unlike Butler, he was not a beekeeper. Instead, he relied on correspondence for his information.

The hive that Wren described (Figure 4.1) was invented by Mr. Mew, the vicar of Eastington in Gloucester, England. Mew gave one of his octagonal hives to Dr. Wilkins, who was the Warden of Wadham College in Oxford. In addition to Wren, Wilkins's hive also attracted the attention of Robert Boyle,

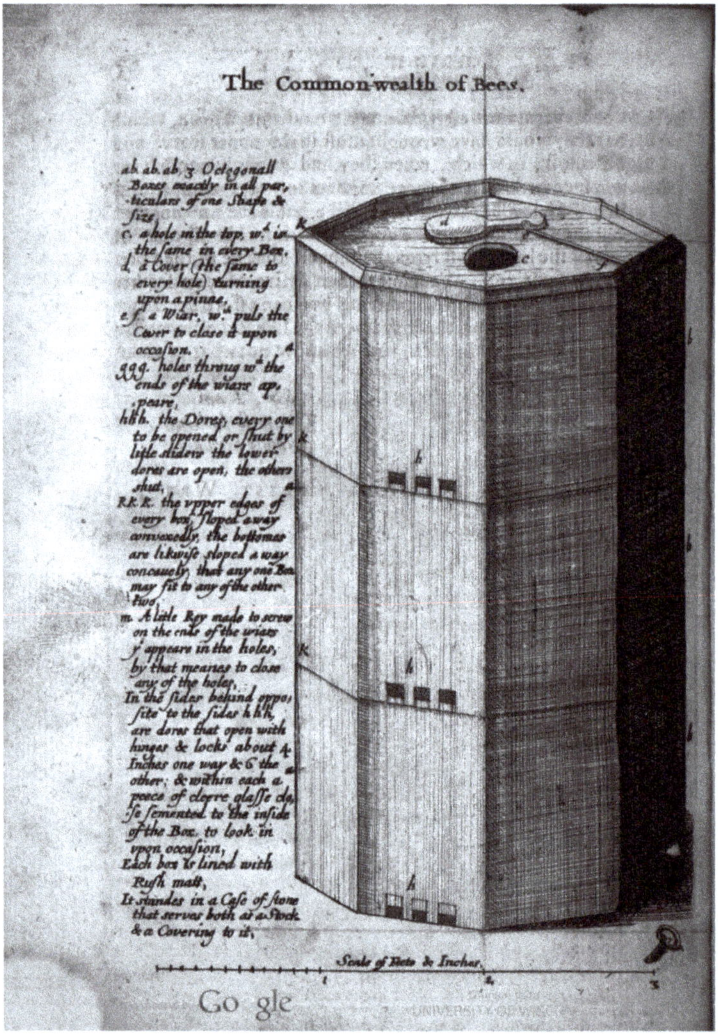

FIGURE 4.1 The octagonal hive described by Christopher Wren (from Hartlib 1655).

considered by some to be the first modern chemist, and Robert Hooke, whose improved microscope permitted Hooke to discover cells (IBRA 1979). It was at Oxford where Wren first saw this new hive, and he sent a description of the hive to Hartlib (1655) in a letter.

Honored Sir.

You have by several hands intimated your Desires to me of having a particular Description of our Three-storied Beehive. I confess I was not over forward to execute this Command of yours; and my Reason was because, the Devise not fully answering our own Expectation, I thought it would be much more unsatisfactory to you: but since you please to persist in your desires (as Mr. Rawlinson told me the other day) I can be no longer shameless to persist in my incivility, especially prompted by mine own Ambition, to find any way to shew myself a Servant to a Person so eminent amongst the ingenious as yourself.

The Description I think is evident enough in the paper; I shall only tell you what effects we find. Last May (as I remember) we put in two swarms together, leaving the places to go in, open only on the lowermost, but all the passage holes open from Box to Box: In the middlemost they first began their Combes then in the lowermost before they had filled the middlemost, & so continued till they had filled both, which before they had quite finished, they began to make 2 little Combes in the upper Box (all this while deserted) & continued besides, a part of a Comb of the middle story, an inch or 2 up into the upper box, filling ~~all~~ almost the passage hole quite up, leaving themselves only a little hole as big as 2 fingers might go in, for their passage up & down: I am not very certain whether this was not done at first when they wrought in the middle Box, & whether this was not the reason why they wrought so little in the upper box, because they stopped themselves up from and easy passage to it: The Combes in the lower stories were well replenished with Honey & suddenly, but these little Combes in the upper they quite desert, contrary to our Expectation, which was that they would have wrought most in the upper story & the middlemost, in which when they had wrought enough for their own spending that then we might take away the uppermost from them, & so have continued still: but if we find another year, that they fill not again the uppermost; 'twill be all one still to take away the lowermost from them; but if that be so then 2 hives will be sufficient. We must rather Desire of you farther light on this business, which I presume you can afford us from other men's observations that have tried the like experiment; for as yet ours is imperfect, & we know not what to make of it. Sir. I am

Your most obedient humble servant
Chr. Wren.

The hive was not very successful, which is generally blamed on the quality of the boards that were thought to be too thin to insulate the bees from the cold winters. A century later, another octagonal hive would be invented in Scotland using thicker boards, which improved the bees' overwintering and produced record amounts of honey (IBRA 1979).

The octagonal hive was patented in 1675 by John Gedde even though it was essentially the same hive presented by Wren. Gedde described it in his book *A new discovery of an excellent method of bee-houses and colonies.* Gedde's hive also spurred an early discussion of ownership of intellectual property. Gedde showed his hive privately to Sir William Thompson, who shared details of Gedde's hive with Sir Robert Murray, who described the hive to the Royal Society and published under Murray's authorship much to the dismay of Gedde (IBRA 1979).

What made Gedde's hive (Figure 4.2) more than a simple octagonal box was the insertion of internal scaffolding to help support the weight of the comb.

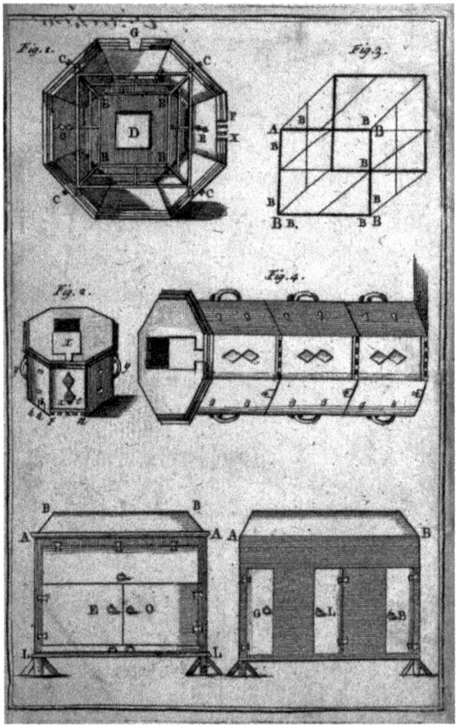

FIGURE 4.2 Gedde's octagonal hive with the internal frame to support the comb (modified from Gedde 1675). Top left the interior of the Gedde hive showing the placement of the frame; Top right the internal frame removed from the hive; middle left a single octagonal box hive; middle right four stacked octagonal boxes.

This "frame" was made of sticks that were fashioned in the shape of a cube with each of the four sides touching the midpoint of alternating boards of the hive (Figure 4.2 top left) (Kritsky 2010).

BARS AND FRAMES

The difficulty with skeps and octagonal hives was that the bees would glue their comb to the walls of the hives. The removal of the comb to harvest the honey was completed using a skep knife or honey cutter that could be inserted down in between the combs to cut away the honeycomb, which was a messy and invasive process. A solution to that issue would be the use of bars or narrow boards, which were placed on top of the hive (Figure 4.3), under which the bees attached their comb. This innovation was first described by Sir George Wheler in 1682. Wheler was not a beekeeper. He was born in 1651 in the Netherlands to parents living in exile during Cromwell's control of England. He was educated at Lincoln College at Oxford, and in 1673 he traveled through France, Switzerland, Italy, and eventually making his way to Greece in 1675.

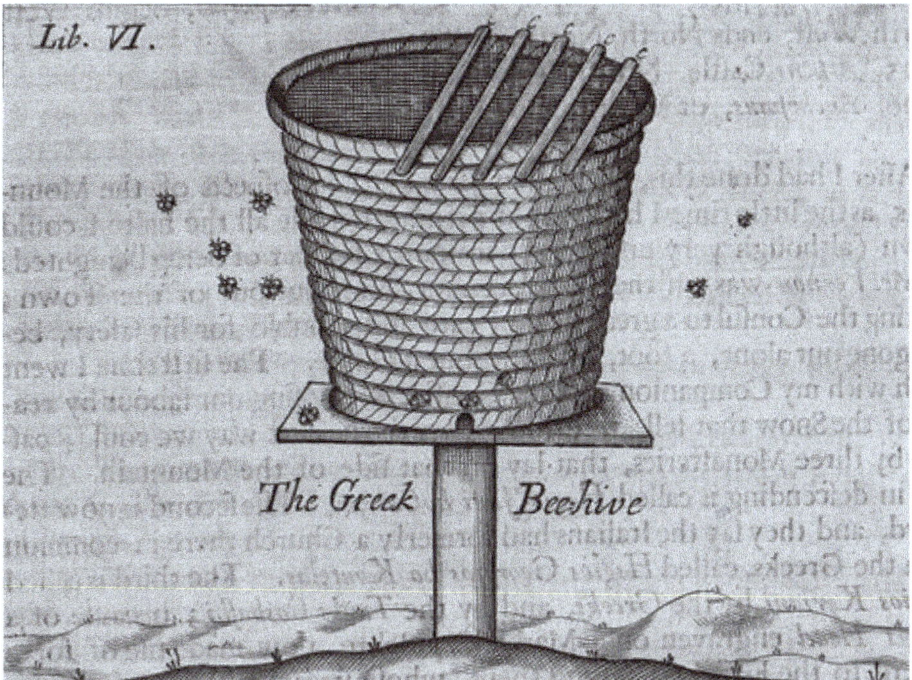

FIGURE 4.3 The illustration of the Greek bar hive from Wheler's book (1682).

While touring Mount Hymettus, east of Athens, he had the occasion to observe an unusual method of keeping bees.

> The Hives they keep their Bees in, are made of Willow, or Osiers, fashioned like our common Dust-Baskets, wide at the Top, and narrow at the Bottom; and plastered with Clay, or Loam, within and without. They are set the wide end upwards, ... The Tops being covered with broad flat Sticks, are also plastered with Clay on the Top; and to secure them from the Weather, they cover them with a Tuft of Straw, as we do. Along each of those sticks, the Bees fasten their Combs; so that a Comb may be taken out whole, Without the least bruising, and with the greatest ease imaginable. To increase them in Springtime, that is, in March or April, until the beginning of May, they divide them; first separating the Sticks; to which the Combs and Bees are fastened, from one another with a Knife: so taking out the first Combs and Bees together, on each side they put them into another Basket, in the same Order as they Were taken out, until they have equally divided them. After this, when they are both again accommodated with Sticks and Plaster, they set the new Basket in the Place of the old one, and the old one in some new Place. And all this they do. in the middle of the day, at such time as the greatest part of the Bees are abroad; who, at their coming home without much difficulty, by this means divide themselves equally. This Device hinders them from swarming, and flying away; In August they take out their Honey, which they do in the day-time also, while they are abroad; the Bees being thereby, they say, disturbed least. At which time they take out the Combs laden with Honey, as before; that is, beginning at each out-side, and so taking away, until they have left only such a quantity of combs in the middle, as they judge will be sufficient to maintain the bees in winter; sweeping those bees, that are on the combs they take out, into the basket again, and again covering it with new sticks and plaster. ... This is done without smoke wherefore the ancient is called is honey, unspoken honey; and I believe the smoke of sulfur, which we use, takes away very much of the fragrancy of the wax; and sure I am, the honey can receive neither good taste, nor good smell of it.
>
> (Wheler 1682)

Wheler included an illustration of this hive in his book, which was the first time a bar hive was brought to the attention of beekeepers in England. Although not realized at the time, the unique shape of the skep, or as Wheler called it, like a dust-basket, was critical in the success of the hive. The tapering of the sides towards the bottom of the hive tricked the bees into thinking that the sides of the hive were actually the bottom of the hive and as such the bees would not attach comb to the sides. This hive would not be fully appreciated for another two centuries (Kritsky 2010).

Wheler's observations on how the honey was taken was also quite novel. In England beekeepers would dig a pit, a little larger than the hive, and provide it with brimstone, or sulfur. This would be ignited, and the beekeeper would shake the hive over the smoldering fire to knock the bees into the fire to kill them. As Wheler noted, this certainly would have tainted the sweet smell of burning a pure beeswax candle.

A year after Wheler's account was published there appeared an account of how to construct a frame to support the comb instead of using a simple bar. The author was identified simply as A.J. The description appeared in a book assembled by John Houghton, an apothecary, who also sold tea, coffee, and chocolate. His book consisted as a collection of letters briefly discussing several topics, with the letter from A.J. accounting for six pages. Such books of letters were popular in the early Enlightenment on everything from book synopses to beekeeping, which helped to satisfy the growing demand for books (Kritsky 2010).

A.J. (1683) did not use an octagonal design for his hive. He instructed his readers to:

> ... take boards in thickness for half an inch, plained on one side, of which make a hive, in the form like onto a foursquare box, observing to turn the planed side of the boards innermost, and to have it of such wood has a smell that offensive to the bees.
>
> (A.J. 1683)

The frames were comprised of a bar that was placed on the top of the box. The sides of the frames extended downward close to the hive walls and closed on the bottom. The author claimed:

> ... when you draw out the frames with the combs and the bees upon them you will find them to come out pretty easy, it were worth the trial, whether would not be less disturbance to the bees (if it can be done) to remove two or three frames at a time, ...
>
> (A.J. 1683)

A.J.'s description was optimistic as the bees tended to build comb between the frames, or if the frames were too close to the hive walls, the bees would glue them to the hive. This was likely why A.J. suggested the beekeeper take two or maybe three frames out at a time when harvesting honey. History will show that A.J.'s promotion of hives furnished with frames was well ahead of its time.

Seventeenth-century books had a limited impact on the public as a whole because the books were expensive for the poor to purchase. The quality of the information was not always accurate, and, once published, a particular book may not receive a wider readership for several decades if at all. This is likely why A.J.'s development of the frame did not lead to more individuals to try spacing frames sooner.

François Huber incorporated frames into a unique observation hive. Huber, who was born in Switzerland, started to lose his vision at age fifteen and was blind as an adult. Yet Huber made critical discoveries about the biology of honey bees with the aid of his wife, son, and assistant. He tried to keep a single comb of bees in a hive, but was concerned that it was not natural for honey bees. Thus, he took a dozen frames and hinged them together on one side permitting the frames to be opened like the leaves or pages of a book. Huber's "leaf hive," when closed was a box with small openings at the lower side of a frame to serve as entrances to the hive. Huber's hive was not designed as a hive for honey production as much as it was a hive to observe and study bees. Although the leaf hive was not a practical hive for beekeeping, it was quite influential in the development of modern beehives (Crane 1999).

AN ENLIGHTENED APPROACH

Not all of the entomological advances of the seventeenth century involved hive design and management. The start of the sixteenth century witnessed new emphases on the biology of honey bees. In particular, the organization of the hive under the direction of a leader or a monarch was a major debate. The bestiaries of the thirteenth century referred to the leader of the hive as a "king." *The Bodleian Bestiary* (Barber 1992) succinctly summed up the widespread belief:

> They fill castles made with wax with their innumerable offspring. They have an army and kings; they fight battles; they flee from smoke and are angered by noise ... They choose a king, create their own people, and all those subject to a king or nonetheless free. For they maintain his right to judgment in there devoutly faithful to him because they recognize him as their elected leader, ... The king is naturally endowed with the special qualities, as his large and handsome body shows, and – the most important virtue that in a king – his friendly nature. He has a sting, but does not use it in revenge.
>
> (Barber 1992)

It was not until the end of the Renaissance that the true sex of the "king" was discovered. Luis Méndez de Torres from Spain was the first to accurately describe the various castes of the hive. A translation of his 1586 book reads:

> To understand what this chapter is deals with, it must be realized that the bee called the [mistress], without coupling with a male and without the pain of childbirth, produces a seed from which are engendered three kinds of bee – mistress, drones and ordinary bees – according to the different cells in which the seed is placed.
>
> (Crane 1999)

Butler (1634) titled his book *The Feminine Monarchie*, but he did not have an accurate understanding of the roles the three kinds of bees played in the hive. Queen Elizabeth I had been his queen of England for most of Butler's life, thus his experience with monarchy was feminine. He also dedicated the 1634 edition of his book to Queen Henrietta Maria, who was the wife of King Charles I (Crane 1999). Indeed, Butler also refers to the queen bee as Sovereign, Prince, and Governor. He does recognize that the drone is a male, but he believed that the drone mated with the workers who produced the next generation. He wrote:

> Reason and sense agreeing, do show that the honeybees are females, but which both sexes, first the females, and then the males are bred. As for the drone it is a gross stingless bee spending his time in gluttony and idleness at others cost ... the truth is, the drone is of the same species with the Honey bee, but are they different sex, ... Every living creature doth generate male and female of its kind, and experience doth yearly teach us, that the bees do breed as well as drones as honeybees, and seeing the honeybees are females, it necessarily followeth that the drones are the males of the same kind.
>
> (Butler 1704)

This uncertainty about the sexes of honeybees highlights the state of apiculture at the end of the seventeenth century. Throughout the first century of the enlightenment, beekeeping literature was mostly written by practical beekeepers well versed in traditional methods. The eighteenth century witnessed greater communication among scientific beekeepers, which led to a better understanding of the biology of honey bees and hive design, and Reverend John Thorley illustrates this change.

Thorley was born in 1671, but started to keep bees around 1704, and when he wrote his major work, *Melisselogia or, the female monarchy* in 1744, he based it on "forty years of observation" (Brown 1994). Thorley settled the debate about the sex of the queen, the role of the drones, designed a new hive, and he discovered some fundamental aspects of honey bee biology (Brown 1994).

Thorley (1744) clearly understood that the queen was a female and provided undeniable observations that were convincing. He wrote of the queen:

> This stately, beautiful, most noble and glorious insect, and so many things expressing Royal Majesty, has been for a great number of years my grand favorite. A princess I greatly admire; very highly value and esteem; and as justly prefer before many thousands of her subjects.
>
> The reason in determining me to give her in this history the title of queen, rather than king, in opposition not only to the ancient, but also some modern writers, the rears will find in this section immediately following ...
>
> The sovereign maybe known from all her subjects by the following criterions, marks of distinction, vis-à-vis by her bulk and stature; by her shape or figure; or by her color and complexion ... the sovereign may be easily distinguished from all of her subjects by her form and shape; being larger and longer than the laboring bees, especially in her hind part, which is by far more taper than all the other terminating in a much sharper point; nature haven't given her this peculiar form, in order the more readily to reach the bottom of the cells, where the eggs are deposited ...
>
> The reasons which have induced me to embrace the ... opinion [that this monarch is] a female, ...
>
> As a further evidence in the case, and what is more fully convincing, ... as I have often proved, placing the insect upon my hand or on a handkerchief, she has (moving upon it) laid several of these oblong [eggs], ...
>
> From the aforesaid remarks, I've taken the liberty to dignify this insect above the rest, and to distinguish her by the name of queen, answerable to the title page.
>
> (Thorley 1744)

Thorley (1744) was not an advocate of killing his bees while taking the honey. Instead he discovered a rather elegant way of drugging the bees using what we call today a puffball, and his instructions sound like they come from a book used by Harry Potter.

> The narcotic, or stupefying potion, is only the fume of the *Fungus maximus*, or larger mushroom, commonly known by the name of bunt, pucksist, or frogcheese; it is as large or larger than a man's head. I had one of these brought to me last summer (unripe and ripe) which weighed some pounds; but when ripe, of a brown color, and turning to powder, they are exceedingly light.

> Shepherd's and herdsman, etc. frequently find them in the fields, and will supply you with them, towards the latter end of the season.
>
> When you have procured one of these puks, put it into a large paper, pressing it down therein to 2/3, or near half the bulk, tying it up very close. Put it into an oven sometime after the household bread is drawn, letting it continue all night ...
>
> When to a whole fire, it is fit for your use in the method following ... With a pair of scissors cut a piece of the puck, as large as a hen's egg (better at first to have too much than too little) and fix it to the end of a small stick, fit for that purpose, and sharpened at the other end, which place so, that I may hang near the middle of an empty hive. This hive you must set with the mouth upwards, near the stock you want to take in a pail or bucket. This is done set fire to the puck with a candle, and immediately place the stock of bees over it, tying a cloth around the hives which you must have in readiness that no smoke may come forth. In a minute's time, or a little more, you will delight to hear [the bees] drop like hail into the empty hive.
>
> (Thorley 1744)

Thorley reported that the bees would wake up in one to two hours, during which time the honeycomb would be cut from the hive and the bees would be returned to their home. Once awake, they would commence to rebuild comb. This "stupefying potion" was also useful in merging two or weak hives. Thorley also used this method to determine how many bees were in a swarm by taking advantage of the anesthetized bees to spread them out and count them on a table.

Thorley (1744) also invented a new beehive that used the best of all hives (Figure 4.4). It consisted of an octagonal box on the bottom reminiscent of Gedde's boxes. The top of the box was fitted with a hole that could be opened or closed with a sliding piece of metal. When the bees had filled up the octagonal box, the hole on the top would be opened upon which a flat-topped straw skep (also with a hole in the top) would be placed. When the skep filled with bees, the plug closing the skep could be removed and a large glass globe would be placed on the skep. Bees would then move into the globe, build comb, and produce honey much to the delight of the observers. Thorley (1744) wrote, "As this receptacle is wholly transparent, the curious observer may entertain himself with reviewing the whole progress of their works." The glass hive had a hole drilled through the knob at the top into which was placed a thick glass rod fashioned with two holes that supported two rods perpendicular to the central glass rod and each other. This provided the support for the heavily laden honeycombs, which could be as much as thirty pounds of honey.

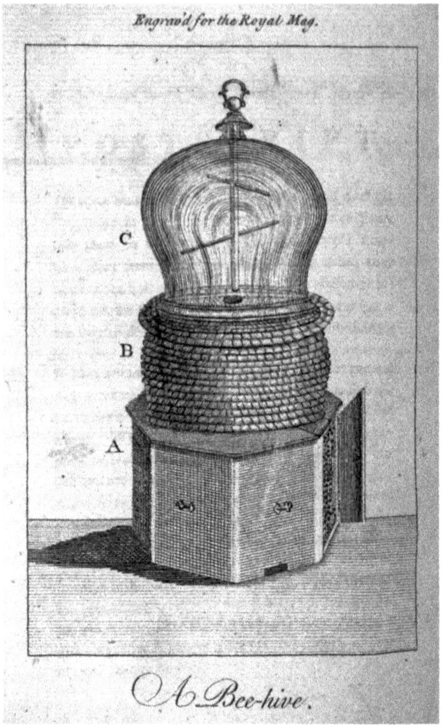

FIGURE 4.4 Thorley's hive consisted of an octagonal wooden hive furnished with windows that were accessed by a door (from Thorley 1744).

In addition to beehives, Thorley recommended the use of a bee-house to protect the hives from the ravages of winter. Bee niches, or bee boles, were first created during the medieval period around 1250, and they continued to be built until 1900. Many of these niches were simple structures, but during the Enlightenment some were quite elaborate. A set of thirty bee boles constructed around 1666 were built into the brick garden wall at Packwood House (Kritsky 2010). Bee-houses, unlike simple bee boles, were often ornate structures that were designed to better decorate the garden (Kritsky 2010). Thorley's bee-house held octagonal hives and was appointed with elegant curves to the roof and rested on legs to keep the base of the bee house dry. Although Thorley (1744) discussed using rather expensive octagonal hives kept in an equally expensive bee house, he also used straw skeps.

Two years after Thorley's book appeared, a new hive, a collateral hive, was advocated by Stephen White. White did not use an octagonal shape; he argued that Gedde's octagonal hives appeared "very expensive and unwieldy" (White 1759). White, instead, used a square box, similar to that proposed by A.J. in

Hartlib's book. White's novel idea was to place these boxes next to each other to permit the bees to walk laterally from box to box, rather than walking up or down within the hive. To give the bees access to adjacent boxes, the sides with two passageways lined up with the neighboring hive. A sheet of tin was slid down between the boxes to close them off from one another. A swarm could be placed in one box, and once comb was constructed, the tin divider would be removed enabling the bees to start working in the adjacent box. White also had windows with shutters built into the hives, permitting him to open the window to monitor how the bees were doing without disturbing them too much.

Four basic hives (wicker skeps, straw skeps, octagonal wood hives, and box hives) were being advocated during the latter half of the eighteenth century. This hive competition changed the bee literature from basic descriptive works to include more critical evaluations of the successes and failures of the various hives. These evaluations also led to more critical observations of honey bees, which increased fundamental knowledge about honey bees. One of the more successful of these books was *A Treatise on the Management of Bees* by Thomas Wildman (1770).

Wildmam was a promoter of beekeeping by staging demonstrations for King George III as well as the general public. He performed with bee beards by strapping a queen bee in a small cage on his chin that caused the swarm to cover his chin and chest like a beard. He would make swarms fly after him while he rode a horse inside a circus ring. His popularity with these demonstrations resulted in nearly five hundred people buying his book in advance of its printing. Despite all this fanfare, Wildman was a forward-thinking beekeeper (Brown 1994). He advocated not killing bees in order to harvest the honey by writing:

> Were we to kill the hen for her egg, the calf for her milk, or the sheep for the fleece it bears, everyone would instantly see how much we should act contrary to your own interest: and yet this is practiced every year in our in human way and impolitic slaughter of the bees. Would it not argue more wisdom in us, to be contented with taking away only a portion of the wax and honey, as is the practice of many countries.
>
> (Wildman 1770)

Unlike other bee book authors, Wildman did not recommend the use of wooden hives, rather he preferred modified straw skeps (Figure 4.5):

> Straw hives, as far as regards to bees, are preferable to any other hesitations, because the straw is not so liable to be heated by the rays of the sun at noon, to which they are generally exposed, and is a better security against the cold, than any kind of wood or other material. Their cheapness renders them an

FIGURE 4.5 Wildman's hive consisted of one or more shallow skeps that would be protected by a straw hackle (from Wildman 1770).

> easy purchase, even to the cottager. I propose that the management of bees in hives shall be altered from what is now practiced, so the size and four of my hives are different from those now in common use ...
>
> (Wildman 1770)

The common skep in use at the time was the bell-shaped skep, and Wildman (1770) figured out an ingenious way of using skeps to simulate a hollow tree, which also was the advantage of octagonal hives:

> My hives are 7 inches in height and 10 inches in width. The sides are upright, so the top and the bottom are of the same diameter a hive holds nearly a pack. In the upper roll of straw there is a hoop about half an inch in breath to which are nailed five bars of pine ... a narrow short bar is nailed to each side, about a half an inch distant for the bars next to them in order to fill up the remaining part of the circle; so that there are in all seven bars of pine, which the bees fix their combs ... When the hives are made, a piece of wood should be worked into the lower role of straw, long enough to allow of a door for the bees of 4 inches in length, and half an inch in height ...
>
> (Wildman 1770)

Wildman (1770) recommended coating the outside of the hive with cow dung as it was "least apt to crack" compared to any kind of cement. The completed hive would be further covered with a straw hackle that rested on top of the hive with the straw extending beyond the hive to help shed the water off of the hive and not wetting the sides. He placed the completed hive on a stool that was made of a single post that was two and a half feet high. He placed soot mixed with barley chaff at the base of the post to keep vermin from crawling up the post.

Wildman would catch a swarm and place it in one of the skeps with the cover. Once the bees started to build comb, he would place a second skep under the first skep. Wildman argued that the queen would lay her eggs in the lowest most skep once the upper skeps were filled with comb. If she had laid any eggs in the upper skep, they would have completed development by the time she had filled the bottom skep. Once these two skeps were crowded, he would place a third skep below the others. When the top skep was filled with honey, it was removed, and a fourth skep would be placed under the third, and so on. His continued success was tied to his leaving about twenty pounds of honey in the hive for the bees to use over winter (Wildman 1770).

Wildman and his son sold octagonal wood hives, but he felt that these hives were more ornamental than practical. He also was critical of White's collateral hives noting that the queens would "lay their eggs promiscuously in both boxes" and therefore the hive did not succeed to do what White had intended. Wildman also felt that Thorley's narcotic to quiet the bees when harvesting the honey was no longer necessary when using his stacking skeps system (Brown 1994).

While Thorley and Wildman were promoting beekeeping in England, Anton Janscha was promoting a very different form of beekeeping in Austria. Janscha was born to a family known for painting detailed beehive fronts that adorn traditional Slovenian beehives and for generations of fine beekeepers. Anton learned beekeeping from his father, who learned from his father, giving Janscha the benefit of decades of experience and confidence. Like Wildman, Janscha was a capable lecturer and eventually was appointed the Beekeeping Instructor at the College of Beekeeping in Vienna. His expertise was recognized by Maria Theresa, the Empress of Austria, who appointed Janscha as the Imperial Beekeeper with a salary, and as such was expected to give lectures and demonstrations to help encourage beekeeping (Brown 1994; Crane 1999).

Janscha used wooden hives that were long rectangular boxes, in contrast to the hives used in England and France. The end boards of Janscha's hives were not nailed into the sides, rather they could be slid inward to accommodate a small swarm or slid outward to provide more room as the bees' population increased (Brown 1994). There were two holes on the top and bottom of the hive that could be opened to provide access to a second box placed above or below the filled hive. When both boxes were filled with bees, Janscha would close off the two boxes and introduce a new queen into the box that didn't have one. Janscha was considered an expert on swarming, and his book, *Handling bee Swarms*, was considered to be one of the best written on the topic (Fraser 1951).

UNDERSTANDING BEESWAX AND THE DEVELOPMENT OF APIOLOGY

The discovery of the origin of beeswax is a prime example of how the ever-increasing number of books inspired more scientific inquiry. The ancient Egyptians thought that beeswax was gathered from flowers (Kritsky 2015), a belief shared by Butler in 1609. The first person to notice wax scales on the underside of the honey bees' abdomen (Figure 4.6) was Martin John in 1684. John assumed that the bees collected the wax as the bees collected honey (Crane 1999). Unfortunately, John's account was published in German and did not make its way to England.

Thorley was the first English beekeeper to observe that the bees carried wax on the underside of the abdomen. His eureka moment still relates his excitement:

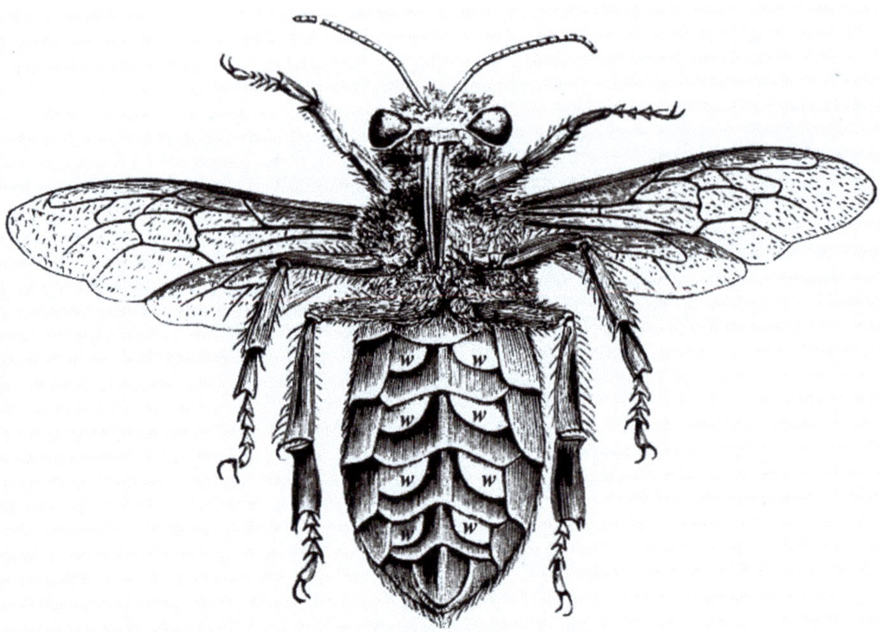

FIGURE 4.6 The underside of a worker honey bee showing the eight white wax flakes extending from between the segments (from Cheshire 1886).

> At last, viewing a hive of bees very busy at labour, I observed one bee among the rest, as she fixed upon alighting place, having an unusual appearance; upon which I seized her directly, before she had time to enter the hive; where, with a very sensible pleasure, I found where I had (till then) been in vain searching for. Upon the belly of this bee, within the plates, were fixed no less than six pieces of solid wax, perfectly white and transparent like gum; three upon one side, and three upon another, appearing to the eye equal in bulk and gravity; so the body of the bee seemed duly poised, and the flight not in the least obstructed by any inequalities. ... once I took away eight pieces all together and I knew that it was wax, and nothing else.
>
> (Thorley 1744)

Wildman read Thorley's description of the flakes of wax on the abdomen, but did not acknowledge it as wax. Wildman confused the sticky nature of the pollen balls brought back to the hive with beeswax. He included a drawing of the hindleg of a worker honey bee describing the pollen ball as "loaded with wax." He wrote that bees "go and collect wax on flowers." Wildman thought the bees collected two forms of wax. He mistook propolis as a brown wax that was used to fill in spaces in the hive, whereas "common wax" was used to construct combs.

The discovery of how beeswax was made required more than casual observation. It required a scientific approach, and the scientist who next turned his attention to the beeswax question was the eminent surgeon and scientist Sir John Hunter.

Hunter watched bees as a teenager, but in later in life he approached honey bees in a very intense way. Paget (1897), Hunter's nineteenth-century biographer, noted that Hunter studied bees for twenty years using a glass observation hive. Hunter believed in the power of observation and experimentation. Whereas Thorley would kill swarms and count the bees, Hunter would kill entire hives, and carefully examine every individual bee in the hive. Given the number of hives Hunter examined, he must have carefully looked at upwards of fifteen thousand individual bees (Paget 1897).

Hunter conducted various experiments on the bees. He would weigh the hives and found they became lighter during colder rainy periods than during the warmer days. He took thermometer readings of hives and found that when cold the bees clustered to keep their hive temperature warm.

Hunter (1792) also observed the bees waggle dance in his hives.

> We very often see some of the bees wagging their bellies as if tickled, running around and to and fro for only a little way, followed by one or two other

> bees, as if examining them. I conceived they were probably shaking out the scales of wax, and that the others were ready upon the watch to catch them, but I could not absolutely determine what they did.
>
> (qtd. in Dobson 1969)

Had Hunter marked the bees to determine they were foragers and followed their flights, he might have discovered the true nature of the dance.

Hunter verified Thorley's observation that the bees had wax scales on the underside of the abdomen and hypothesized that the bees secreted the wax. He provided support for his hypothesis with an experiment and further observation. He reported his conclusions to the Royal Society of London in 1792:

> The wax is formed by the bees themselves; it may be called an external secretion of oil, and I have found that it is formed between each scale of the underside of the belly. When I first observed this substance, in my examination of the working bee, I was at a loss to say what it was: I asked myself if it was new scales forming, and whether they cast the old, as the lobster, &c. does? but it was to be found only between the scales, on the lower side of the belly. On examining the bees through glass hives, while they were climbing up the glass, I could see that most of them had this substance, for it looked as if the lower, or posterior edge of the scale, was double, or that there were double scales; but I perceived it was loose, not attached. Finding that the substance brought in on their legs was farina, intended, as appeared from every circumstance, to be the food of the maggot, and not to make wax; and not having yet perceived anything that could give me the least idea of wax; I conceived these scales might be it, at least I thought it necessary to investigate them. I therefore took several on the point of a needle, and held them to a candle, where they melted, and immediately formed themselves into a round globe; upon which I no longer doubted but this was the wax, which opinion was confirmed to me by not finding those scales but in the building season. In the bottom of the hive we see a good many of the scales lying loose, some pretty perfect, others in pieces.
>
> (Hunter 1792)

Although Hunter correctly discovered that wax was a secretion, he did not believe that the bees could produce enough of it to account for all the honeycomb in the hive. He published in 1792 that the bees also needed pollen to produce wax. Hunter's paper was translated by Huber, who conducted a series of experiments to determine if pollen was required by the bees to make beeswax. Huber placed a swarm of bees in a hive in a dark cellar and provided the swarm with water and honey. The bees were allowed to fly about the cellar, but were not permitted to forage. After two days the hive was examined and found to have five new combs made of white wax. Huber repeated his experiment twice more with the same result. He then fed the bees only pollen and water for eight days and discovered the bees had constructed no honeycomb (Brown 1994). Huber established that honey bees need honey to make wax.

The origin of wax illustrates how accumulated knowledge during the Enlightenment was changing our understanding of bees. Butler espoused the views of the ancients that bees collected wax from plants. John and Thorley observed wax flakes on the underside of honey bees' abdomens. Observation is the first step to what is now known as the scientific method. These observations lead to hypotheses, which can then be tested, as both Hunter and Huber conducted experiments to reveal the truth about how beeswax is made. Hunter and Huber were interested in bees, not necessarily beekeeping, and that is the difference between apiculture, or beekeeping, from apiology, the scientific study of bees.

APICULTURAL SOCIETIES

The growing availability of beekeeping knowledge also promoted the formation of beekeeping societies. Many of the early societies, such as the Royal Society founded in 1660 and the Académie Royale des Sciences founded six years later, were for the elite and were broad in scope. However, as interests became more specialized so did the societies. The German *Bienengesellschaft* was founded in 1766 by leading beekeepers. Their mission was to promote beekeeping by teaching, mentoring, and creating a dialogue on improving techniques. Soon other German beekeeping societies formed in Rotha, Bavaria, and Munich (Crane 1999).

British beekeeping societies started forming after 1796 when J.C. Lettsom published a small pamphlet titled *Hints for promoting a Bee Society.* Four such societies were established in Britain by 1799 with the goal of promoting economic opportunities through beekeeping. Reverend J. Issac wrote a small 128-page book titled *The General Apiarian* with a subtitle that read "a simple, humane and advantageous method of obtaining the produce of bees without destroying them." The proceeds of the book helped fund the Western Apiarian Society. These early societies paved the way for the beekeeping associations that appeared during the Industrial Age (Crane 1999; Kritsky 2010).

THE IMPACT OF BEEKEEPING DURING THE ENLIGHTENMENT

The Enlightenment witnessed what we now realize was a revolution in obtaining the bees' produce, honey and wax. At the start, the common skep hive, woven from straw or wicker, was the hive of choice for beekeepers. However, the availability of bee books with novel hive designs provided beekeepers with the incentive to experiment with new hives. The skep remained popular throughout this period because they were inexpensive, lightweight, and their use was almost general knowledge. On the other hand, wooden box hives were expensive. Thus, these new wooden hives were found mostly in formal gardens owned by the landed gentry.

Beekeeping from 1600 to 1820 laid the groundwork for what would be known as "rationale beekeeping," based upon "thought and reason rather than inherited tradition" (Crane 1999). The invention of novel hives that were promoted through printed books galvanized beekeepers to experiment, which in turn led to new developments in beekeeping.

Eva Crane (1999) wrote that there were five major innovations that led to the modern rationale of beekeeping: the use of rectangular boxes, tight-fitting boxes placed on top of each other, using bars under which the bees could build comb, enclosing the bars on all sides to frame the comb, and the spacing of the frames to make them freely moveable. All but the spacing of the frames were present by 1683 (Kritsky 2018). The spacing of frames would not be figured out until the Industrial Age, but the first four of the important innovations revealed themselves to enlightened beekeepers who were not afraid to experiment. This foundation of knowledge, combined with the freedom to experiment that was promoted through printed books, would go on to change beekeeping within the first five decades of the Industrial Age.

CHAPTER FIVE

Insects in Mythology and Religion

DAYTONA D. JOHNSON AND ELLEN M. CAMERATO

As the world transitioned into the Age of Enlightenment (1600–1820), also known as the Age of Reason, many cultures moved away from holding myths as the center of their belief systems. Some groups, however, still held onto folklore as a form of entertainment and as guidelines for their daily practices. Insects have always been a significant part of human life, carrying on various roles from pollinators to pests to food sources, and many people derive pleasure from the corpus of science and art (Southwood 1977). It is not surprising, then, that their prevalence has also allowed them to creep their way into the belief systems of cultures around the world.

The observations of insects by professionals and amateurs allowed the scientists of societies during the Age of Enlightenment to reflect on oneself as well as the moral and societal issues of the time (Coleman 2006). The study of how insects and other arthropods fit into religion and mythology falls into the realm of cultural anthropology and describes a story from sometime in the past that is used to explain and represent cosmological and supernatural traditions of people and their cultural traits, as well as religions (Cherry 2004). Mythologies are notoriously deep-rooted and resilient beliefs held by cultures through time (Daston 1998). The focus of this chapter is to examine how the incorporation of anti-Christian sentiment influenced religious beliefs during the Age of Enlightenment and how mythological remnants of ancient cultures involving insects shaped society and belief systems during that time period.

Although the Age of Enlightenment has some semi-established parameters and characterizations, such as reason and rationalism, the accounts and definitions of the time period varied tremendously throughout the world. Johann Friedrich Zollner, a pastor in Berlin, wrote the article "What is Enlightenment?," which is a question we are still asking today (Outram 2019). In a narrow sense, the Age of Enlightenment can be described as a period of time that was characterized by the emergence of freethinkers who were consumed with an insatiable hunger for knowledge and reason rather than superstition based on conventional ideals (Outram 2019). This new attitude, an opposition of traditional social thinking and blatant criticism of established religion, especially Christianity, was particularly prevalent in parts of Europe. Admittedly, the Age of Enlightenment is too amorphous of a concept to adhere to specific guidelines or definitions and different countries adopted these ideas and concepts at different times (Byrne 1997).

BUILDING ON THE RUNES OF REFORMATION

While this era is often referred to as the Age of Reason, as if reason did not exist before this movement, it should be noted that it was also a time of mysticism and exploring the natural world in unconventional ways. Some projects, such as mesmerism, erupted during this time due to the hard push against established religion and the embracement of the natural world (Byrne 1997). Developed by Franz Mesmer, mesmerism was the belief that all living things possessed a natural invisible force that contributed to the overall wellbeing of the individual, which has since been deemed a pseudoscience. Mesmerism, however, eventually led to the development of hypnotism, which suggests that one can slip into a trance through the power of suggestion rather than relying on some form of occult mesmeric phenomena (Byrne 1997).

The Age of Enlightenment was built on the runes of the Reformation. Europe was in a state of crisis from 1490–1648, which included events such as the fall of Constantinople and the religious upheavals that followed. For the first time in history, the Bible became widely available, and due to an anxious climate many people held it close. Christ's words regarding the Last Days was frequently quoted during sermons. Between 1498 and 1650, at least 750 separate editions of the book of Revelations were printed and of the fifteen illustrations created by Albrecht Dürer for an edition (produced in both German and Latin languages), the piece that depicts the Four Horseman of the apocalypse is the most famous (Cunningham and Grell 2010).

The text represented in Durer's work is as follows:

> And I saw when the Lamb opened on of the seals, and I heard, as it were the noise of thunder, one of the four beasts saying, Come and see.
> And I saw, behold a white horse: and he that sat on him had a bow; and a crown was given unto him: and he went forth conquering, and to conquer.
> And when he had opened the second seal, I heard the second beast say, come and see.
> And there went out another horsè that was red: and power was given to him that sat thereon to take peace from earth, and that they should kill one another: and there was given unto him a great sword.
> And when he had opened the third seal, I heard the third beast say, Come and see. And I beheld, and lo a black horse; and he that sat on him had a pair of balances in hand.
> And I heard a voice in the midst of the four beasts say, A measure of wheat for a penny, and three measures of barley for a penny; and see thou hurt not the oil and the wine.
> And when he had opened the fourth seal, I heard the voice of the fourth beast say, Come and see.
> And I looked, and behold a pale horse: and his name that sat on him was Death, and Hell followed with him. And power was given unto them over the fourth part of earth, to kill with sword, and with hunger, and with death, and with the beasts of the earth.

This illustration portrayed the curses of war, famine, disease, and death. Europe quite literally experienced all of the *horseman*, which left society in shambles. False prophets, wars, and famines plagued Europe for years and people truly believed that Revelation was on the cusp of coming to fruition (Cunningham and Grell 2010). The Great Famine and Black death that caused millions of deaths occurred only about 150 years prior (Nelson n.d.), and one could have speculated that pestilence, famine, war, and death had made their arrival on Earth. Pestilence, in particular, was blamed for the Black Death (1347–1351), which was facilitated by insects (fleas). The plague killed at least fifty million people in Europe (Nelson n.d.). It was common for people during this time to sing songs about the plague, perhaps to lift their spirits, while they believed they were being persecuted by pestilence. A common song sung during the Black Death:

> "A sickly season," the merchant said,
> "The town I left was filled with dead,
> and everywhere these queer red flies

crawled upon the corpses' eyes,
eating them away."
"Fair make you sick," the merchant said,
"They crawled upon the wine and bread.
Pale priests with oil and books,
bulging eyes and crazy looks,
dropping like the flies."
"I had to laugh," the merchant said,
"The doctors purged, and dosed, and bled;
"And proved through solemn disputation
"The cause lay in some constellation.
"Then they began to die."
"First they sneezed," the merchant said,
"And then they turned the brightest red,
Begged for water, then fell back.
With bulging eyes and face turned black,
they waited for the flies."
"I came away," the merchant said,
"You can't do business with the dead.
"So I've come here to ply my trade.
"You'll find this to be a fine brocade ..."
And then he sneezed.

While we may not see the direct impact of insects on medieval society from a religious or mythological perspective during the Dark Ages, we can actively observe through the analytical observation of historical text that insects played a large role in the death and decay process during that time that surely stayed with society. Flies were not associated with anything *good* or *clean* and although society did not have the knowledge or understanding of disease transmission, it is apparent that they associated flies with death and decomposition. This association likely fueled a fear and disgust for the insect. This fear was coupled with pestilence and the Black Death and everything associated with it led to the weakening of the church. This eventually would indirectly give rise to the Protestant Reformation and eventually the Enlightenment.

The Black Death, although not immediately, did impact Europe politically through the weakening of the church. The public genuinely believed they were being punished by God and when the church could not explain why they were being punished or offer a cure they became bitter and the church lost tens of thousands of its congregation from the plague, and from those who lost their faith in the church along the way, this would be the foundation for the Protestant Reformation. Without the Protestant Reformation, there may have never been dissatisfaction with the Catholic Church, and Protestants may have never left for the New World. The Black Death was one of the most tragic epidemics in European history, but it was also the most defining (Peschke 2007).

Structured religion that requires faithful congregations to adhere to a strict guideline usually produce the idea that any deviation from the *normal* and *faithful* path results in heresy. The word "heresy" originated from Judeo-Christian tradition and derived from the Greek word *hairesis*, which means "deliberate choice," and is understood to be an error in lifestyle that does not agree with one's religious commitments. During the reformation, the confessional and institutional unity of the church began to crumble, and different sects of the Christian faith began to use heresy as an all-purpose term for religious mockery. Those of different afflictions began to throw accusations at one another—Catholics against Protestants and even arguing amongst their own faiths (Laursen et al. 2005).

It is challenging to comprehend early modern English religious life without reference to the Puritans. As a group, they campaigned against the hierarchical nature of Anglicanism and promoted the Presbyterian non-hierarchical Church, which was to be the pure apostolic church. The goal of the puritans was to ultimately create an independent church that posed a threat to the Tudor and Stuart monarchs during that time. When the Puritan church first started, Queen Elizabeth put an end to it, which eventually led to the English Civil Wars (Barnett 2013). The Puritan Reformation alongside the Enlightenment led to secularization, often referred to as the rationalization of the modern world facilitated by the disenchantment of the universe. This did not mean the absolution of magic from the world, but rather the separation of magic from religion in early modern Europe.

During the Pre-Reformation, magic in Europe was closely intertwined with religion and the Reformation took all magical elements out of Christianity through the elimination of ritual magic, the idea that any object held some sort of supernatural power, and that human actions were not supernatural at all (Scribner 1993). Two events made the Age of Enlightenment possible: the first being the glorious Revolution (1688), which established constitutional limits on monarchical power and granted religious toleration, and the second Sir Isaac Newton's publication of *The Mathematical Principles of Natural Philosophy*, which stated that nature was orderly and governed by cycles of the universe. Nature was the foundation for the Age of Enlightenment thinkers, but Paganism was likely an inspiration.

HERETIC ROOTS OF THE OLD WORLD

During the Enlightenment there was a characteristic widespread romanticism and interest in the origins, nature, and meaning of myths from the ancient worlds. This interest was especially strong in England where there was an undeniable desire to resurrect divinities that had been long forgotten. The purpose of this exploration was to understand the natural world and give modern meaning and interpretation to ancient mythology. Mythologies are notoriously deep-rooted and resilient beliefs held by cultures throughout time (Daston 1998). This exploration is motivated by theological considerations, and due to the state of Europe after the Reformation, many scholars hoped to establish a unified basis for which they rested their faith by trying to prove that the heathen gods were in fact still Christian figures going by a different name. For example, the theologian and philosopher Pierre Daniel Huet stated that the heathen gods were actually Moses in order to make his Christian faith more established (Kuhn 1956).

It was the Romantics who rediscovered Homer, Plato, Aeschylus, Virgil, and the myths of Prometheus. Above all, philosophers lacked a Dionysian feeling as opposed to the Apollonian elements of pagan antiquity, but more than science separated the philosophers from their classical ancestors (Leith 1971). The constant struggle between the established Christian faith and the emergence of recycled paganistic ideas, coupled with scientific exploration, created a new world. A world where humankind no longer turned to a faceless god, but instead threw themselves at reason and nature. They listened closely and even paid tribute to the heretic roots that were established long before Christianity.

Indeed, small pieces and parts of an older world slowly embedded itself into the Age of Enlightenment. Latin words, for example, used by the Romans to describe ghastly malevolent spirits ended up being recycled and used to describe groups of organisms by Carolus Linnaeus. The Romans believed that when a man died, he may be punished by becoming a larva. Larvae were believed to be demonic spirits that were once human and as repercussion for their wrongdoings committed in life they were punished and sentenced to an eternity of hopeless wandering and exile. While these spirits were believed to be malevolent, it was said that they would not harm good men but would harass and abuse evil men (Guiley 2009). The etymology of the word larva can be traced all the way back to 1600 and meant "mask" in Latin. The Oxford dictionary states that in the mid-seventeenth century the word denoted a disembodied spirit or ghost.

The Romans even observed the festival of Lemuria to appease these malevolent spirits and exorcise them from households to prevent them from causing harm or trouble to its inhabitants. The most sacred ritual was executed by the head of the household on the night of the festival. The owner of the home would carefully wash their hands three times, place black beans in their mouth, make the sign of horns with their hands, and would then toss the beans over their shoulder—making sure to never look behind them. While tossing the beans they would chant, "With these beans I do redeem me and mine." The incantation would be repeated nine times and it was believed that the evil larva would pick up the beans and leave until the following year (Guiley

2009). Carolus Linnaeus even used the word larva to describe immature insects that did not resemble their adult counterparts, and thus *masked* their identity. It also inspired Linnaeus's coining the word lemur to describe a lineage of primates. It has been incorrectly assumed that he chose this name based on their masked appearance, whereas he actually chose the name because of the ghostly reflective eyes and ghastly screams emitted by lemurs (Dunkle et al. 2011–2012).

The Enlightenment was a monumental period in intellectual history that dismantled established beliefs and gave rise to anti-religious sentiment; however, the rejection of Christianity did not result in the rejection of all forms of religion. The skepticism and criticism of organized religion centered around traditions and institutions of the past, such as the church. During this time, individuals greatly admired the artistic and cultural achievements of Greece and Rome. This exploration and resurrection of other cultures and beliefs led to the adoption of pagan virtues as useful guides that would still lead to heaven. The discovery of new lands and scientific advances started to suggest that the European Christian was not the center of the universe nor the only group of individuals that play a part in a divine plan. This mindset was the start of society viewing themselves as a part of something greater and the first challenger of the European anthropomorphic mindset (Byrne 1997).

The importance of insects in religion extended beyond Europe during the Age of Enlightenment. For example, the Indian subcontinent, under the influence of Hindu and Buddhist traditions, cherished and respected insects because of the belief in reincarnation. A major facet of Jainism is to avoid causing harm to any living thing, which extends to insects. Jains are exclusively vegetarian to accommodate this principle. Directly related to insects, Jains may take on several practices to avoid harming them, including drinking water only in the light to avoid swallowing an overlooked insect, wearing a mask to prevent inhalation of smaller insects, and walking with a broom to sweep insects aside before stepping on them. In more strict sects, like the Digambara monks, followers do not wear any clothes and just carry a peacock feather to sweep insects aside, or some even take vows to limit the number of steps they take each day. The Muslim Mogul emperors adopted these attitudes, and on the grave of Nur Jahan (1577–1645) the following was written (Southwood 1977):

Do not decorate my grave with
flowers or candles
Because the flowers may deceive
The song birds
And the candle may burn the
Moth

CHRISTIANITY, WITCHCRAFT, AND DEMONIC INSECTS

In medieval times, Beelzebub was thought to be a demon with great power who ruled over witch sabbats. These sabbats involved denying Christ and mystical orgies that were rumored to lead to epilepsy, apoplexy, or even strangulation. The term sabbat for these mystical parties and meetings was not used consistently until the mid-fifteenth century. The sabbat played a more prominent role in rest of Europe rather than England until 1612, when the Lancashire witch trials took place. Witches were also accused of holding Black Masses, although no single characteristic can be used to describe these events, they were believed to be a parody of the Catholic Holy Mass at which the Devil and his counterparts were worshipped. Beelzebub's name was often chanted during these meetings; perhaps even more frequently than the Devil's (Guiley 2009).

Beelzebub was revered as a formidable demon. Commonly referred to as the "Lord of the Flies," it was believed that he would manifest as a giant fly or truly monstrous being. Beelzebub was deeply feared and was said to have been the Prince of Demons by the Hebrews, around the time of Jesus. The Pharisees even accused Jesus of exorcising demons in the name of Beelzebub, and in the Testament of Solomon, Beelzebub is controlled by King Solomon through the use of a magical ring that rendered the demon powerless. It was believed that Solomon used Beelzebub to interrogate other demons in order to gain information about heavenly affairs. After the crucifixion of Jesus, Satan told Beelzebub that he was going to drag Jesus to hell to get revenge for all the times that Jesus escaped and overpowered him. When Jesus arrived, the Lord of the Flies cast Satan from the mouth of hell and barricaded the gates with heavy chains in attempt to keep Jesus from entering. To no avail, Christ broke through the chains and freed all the saints that had been dragged to hell to be tormented. After Jesus leaves, Beelzebub took the throne and became the supreme ruler of hell (Guiley 2009).

Beelzebub was blamed for the misfortunes in the Loudun Possessions (1630–1634). The Loudun Possessions are one of the most famous cases of suspected mass demonic possession. This series of events took place in Loudun, France where the Ursuline Convent found themselves bewitched by Father Urbain Grandier. Grandier was said to have been a handsome and wealthy man; however, he was incredibly arrogant. After cultivating many enemies, and the convent convincing other priests that Grandier had caused their possessions, torment, and hallucinations it was an easy feat for his enemies to condemn him. A grand total of twenty-seven nuns were accused of being possessed and the exorcisms that took place as part of the trial quickly became a source of entertainment. It was apparent during the trial that the nuns were lying about the possessions. Despite the lies, the case against Grandier was upheld. In fact, the *possessed* nuns even testified as the suspected demons that plagued them. One of the most interesting pieces of evidence that surfaced during the trial was a pact made between the Devil and Gradier that was allegedly stolen from Lucifer's cabinet of devilish agreements. This agreement was signed by Satan, Beelzebub, Lucifer, Elmi, Leviathan, and Astaroth. On August 18, 1634 the Royal Commission passed sentence and declared that after the first and last degrees of torture, Grandier would be burned alive. It was reported that while Grandier was being burned alive, a large fly buzzed around his head, which some speculated to be Beelzebub preparing to take Grandier's soul to hell (Guiley 2009).

During the Enlightenment, the diagnosis of possession was expanded to include all disturbances believed to manifest in the person's afflicted soul. The widening of the diagnosis was in response to the rapid growth of ecstatic behaviors that were engulfing society. While the diagnosis of possession mostly led to pain and agony, it did come with the hope that there would be a diagnosis or a meaning to someone's suffering, which might be why individuals claimed to be possessed by malevolent spirits rather than the divine. Due to the lack of medical knowledge at that time, almost all afflictions were categorized as possessions by clergyman. This might be due to the Black Death, where clergyman lacked answers and explanations for suffering, which the church later suffered for. It also should be noted that many more woman were affected by this than men, probably because they were believed to be easily tempted by the Devil. Individuals also exploited that movement to attack their enemies, such was the case with Grandier (Sluhovsky 2007).

Mass possession cases, like the one in Loudun, became major theaters of Catholic anti-Protestant propaganda. The Diabolic possession always aligned within a Christian demonological framework that assumed Satan and his malevolent subjects were on a persistent campaign to harm humanity. It was thought that living a truly Christian life could prevent such attacks. There was an increase in witchcraft during the Enlightenment; however, few possession cases resulted in accusations of witchcraft (Sluhovsky 2007). It was stated by Kittredge that the essential element in black witchcraft is *maleficium*, which is the working of harm to the bodies facilitated by strange powers. This belief was embedded into society long before Christianity, before the Enlightenment. It was stated by Brother Francesco Mario Guazzo (1608) that the evil Master of witches instructs them to assemble on Sabbat and infest trees with locusts, caterpillars, slugs, butterflies, cankerworms, and such pestilent vermin that will devour everything. Such insects were often feared by superstitious Christians during this time due to the belief that the Devil was orchestrating these manifestations (Weiss 1930).

The role of insects in witnessed miracles was published in 1687 in a manuscript from the English School of the Eucharist as part of the Catholic

Church. These experiences were meant to be used in devotionals by members of the church. An entire section is dedicated to bees, including a story of a peasant in Auvergne, France who had feared that his bee colonies would perish. He was advised to receive communion then blow the Host, or the bread component of the Eucharist, into the hives. The Host fell on the ground, but the bees emerged from the hives, lifted the Host on their wings and placed it in the combs. However, the peasant then later observed the bees to be dead and saw that the Host had turned into an infant, which he also perceived to be dead. When he took the child to bury it on the church grounds, he realized it had vanished from his hands. In another tale, grasshoppers or locusts were swarming and consuming the people's crops. The abbot in the area commanded the people to pray to God and the insects were later found dead on the ground, thought to have sacrificed themselves to fertilize the soil. This publication also included a story of a person drinking from the consecrated chalice of wine that a spider had fallen into and the spider later emerging alive from the person's skin, mouth, or nose. Caterpillars are also mentioned, taking on the role of deflecting the Devil (Bridoul 1687).

DEISM, NEOPAGANISM, AND FOLKLORE

As some people during the Age of Enlightenment started moving away from Christianity, nature replaced God as the ultimate basis for legitimation. Natural law decided political orders and natural theology proved the existence of the benevolence of the Deity (Daston 1998). Deists are traditionally regarded as radicals who rejected revelation as proof for religion and preferred natural religion expressed through the deep exploration of what lays within reason and consciousness. This outlook naturally produced a critique of the Catholic Church, but still suggested that God created the universe; however, God had no influence after creation. Conservative Christians complained that Anglicanism was undergoing change, and most denounced Deists. Unfortunately for the Deists, they were few in number and studies have faced a shortfall in quantifying their numbers. We do know, however, that although naturalistic writing was important to Deists, many other groups were writing about the natural world during this time. This concept was influential both scientifically and politically. The dominant Christian culture even adapted some philosophical beliefs from nature (Daston 1998).

Jean Jacques Rousseau (1712–1778), a famous Deist philosopher and writer, contrasted nature and civilization with good and evil, which presented a picturesque idea of nature that involved freedom from tyranny and positivity. Rousseau's influence over future writers would shift our whole perspective of nature and insects by meshing Deism with philosophy. Rousseau's writing led to famous pieces, such as *Hurt no Living Thing* from *Sing-Song* by Christina Rossetti (1830–1894) (Kacane 2012). The attitude towards insects is a pleasant one:

Hurt no living thing:
Ladybird, nor butterfly,
Nor moth with dusty wing,
Nor cricket chirping cheerily,
Nor grasshopper so light f leap,
Nor dancing gnat, nor beetle fat,
Nor harmless worms that creep.

The world, according to Rousseau, was a substance created by God where people are a part of the harmonious system of nature. Nature worship typical of the Romanticist writers would be inspired by Rousseau's strong Deist beliefs (Kacane 2012).

Enlightenment thought, articulated by Deists, held on antagonistic views toward Christianity and Judaism; these individuals wanted to do away with these religions. Some Deists, however, equated Christianity and Judaism with natural religions and wanted to remove the unreasonable and corrupt pieces of those religions. Deists believed that their god had shown a light of reason on the world and that a reasonable god would reveal this revolution to all groups of people and not expect them to partake in only one peculiar worship or mysterious doctrine. The Deists as a group would condemn witches who did not believe in the revelation, but also encouraged polytheism (Eilberg-Schwartz 1989).

While the Age of Enlightenment is often believed to be responsible for the rise of neopaganism, pagans and witches believed that the emphasis on reason was just as corrupt as the emphasis on religion (Eilberg-Schwartz 1989). Interestingly, there are some striking similarities between neopaganism and the critique of religion during the Enlightenment. Both Deists and paganists define themselves in opposition to the religious teachings of Christianity and Judaism and hold both religions partially accountable for some of the ills of modern life (Eilberg-Schwartz 1989). Deists believed that the God of Reason created the world but allowed it to operate on its own with no supernatural involvement post-creation. Whereas neopagans believed that human life, nature, and the divine should be merged into one concept (Eilberg-Schwartz 1989). The Religion of Nature was believed to restore dignity to oneself because the individual was no longer dependent on a tradition or the knowledge of what is good or moral. This religion was accessible to all individuals through the practice of reason. No clerics, levels of anointment, or one person could claim exclusive knowledge of the truth within the religion, and spiritual exploration varied greatly from individual to individual (Eilberg-Schwartz 1989).

Folklores were mainly communicated during the Enlightenment in bourgeois households where religion was intertwined with romanticism (de Blécourt 2018). This interwoven system of fantasy and religion was the foundation of oral storytelling and lesson teaching to children for centuries. However, it was during the Age of Enlightenment that two brothers from Germany wrote about these stories, known as the Brothers Grimm. The brothers would eventually be bestowed the task of gathering oral tales from the German culture to honor their heritage. They began their exploration in 1807 and the purpose of their collection was to document the oral traditions of old German *Poesie* and to demonstrate how cultivated literature evolved from traditional folk material.

The brothers' strict Reformed Calvinist upbringing taught them to practice diligence, industry, honesty, and order when retelling the old Germanic tales. For example, *The Queen Bee* by the Brothers Grimm is a story where three brothers set out on an adventure. During their journey they stumble upon an anthill and the two eldest brothers want to destroy it. The ants run away from the hill in terror, but the youngest brother would not permit the brothers to destroy the anthill. The brothers then come across a large number of ducks and the two oldest brothers want to roast them, but yet again the youngest brother will not permit it. Lastly the brothers find a bees' nest and the two oldest brothers want to suffocate the bees with smoke, and the youngest brother will not allow them. They later find a castle to rest in and are given three tasks, and if completed they can remove the castle from enchantment. The first task is to find one thousand pearls lost in the moss bed within the forest. The oldest brothers are unable to complete the task and subsequently are turned to stone, but the youngest brother, with the help of the ants he had previously saved, found all of the pearls. The second task is to fetch a key from the bottom of the lake that would unlock the King's daughters bed chamber, and with the help of the ducks he had saved earlier he was triumphant. The third and final task was the most difficult: the brother was to pick the youngest and sweetest of the King's daughters to marry, but they all looked of similar age. Before the princesses took a nap, they each had a sweet treat and the youngest had eaten a spoonful of honey. While the brother was trying to decide, the Queen Bee he had saved earlier arrived and landed on the lips of the youngest daughter and confirmed that she was the youngest and sweetest. The enchantment was broken, and his brothers returned to their natural forms. The youngest brother went on to marry the youngest daughter and eventually became king (Grimm et al. 2011).

Many cultures have their own versions of stories with lessons similar to those presented in Brothers Grimm, even if the names of the Gods and the religion are different. The Protestant influence on such tales in Germany took hold and was prominent due to the political climate in Europe and the influence of puritan religion in that region of the world (Zipes 2002).

Peter Gay was one of the first men to link the American colonies of Britain to the Age of Enlightenment in a series of books beginning in 1966. He discussed in his synthesis that the commitment to *Life, Liberty, and the Pursuit of Happiness* was the fulfillment of Enlightenment parameters and that the movement had indeed found its way outside of western Europe (Outram 2019). In Gay's publication of *The Rise of Modern Paganism* (1966) he started his analysis of the Enlightenment and would go on to publish a second volume titled *The Science of Freedom* (1969) and finished the series with *The Bridge of Criticism* (1970). Part of his account is organized chronologically to show how the Roman Enlightenment was engulfed by Christian culture until Renaissance humanists and seventeenth-century thinkers paved the way for philosophers. Gay went on in his volumes to cast light on the influence of Lucretius and other pagan thinkers that had a deep connection to nature (Leith 1971).

THE NEW WORLD AND OLD TRADITIONS

In the United States, mythology and folklore were spread as oral traditions. A new nation (from the European perspective) was born in the seventeenth century and the new colonists there possessed no folklore or traditions besides stories from their homelands. In a relatively short amount of time, these new Americans would establish distinct literature, behavior, and folklore. The shaping of this folklore began with the first landings on new soil and drew from two heritages, the uprooted European and the Native American (Dorson 1993).

The European settlers who crossed the Atlantic in the seventeenth century brought with them a suit of supernatural beliefs and ideas of the universe. They still gave credence to demons, hags, ghouls, and all specters of the night. They even regulated their lives with magic charms, sigils, and exorcisms. Supernatural explanations in terms of God and the Devil ruled during this time period and colonists viewed and interpreted the witchcraft, demonology, and divine providences. American folklore and mythology resulted from a grafting of the beliefs of the Old World with the environment of the New World. Throughout the seventeenth century, many settlers gathered to exchange oral stories about the New and Old Worlds. Medieval bestiaries fostered zoological myths in America and contributed greatly to the lore of North America (Dorson 1993).

The Age of Exploration produced interesting stories and traditions (Dorson 1993). For example, colonists were fascinated with the antlion and eventually manifested into speaking charms at them, intending to entice the animal out of the bottom of its funnel. An example of an antlion charm chanted during the Enlightenment era is:

Doodle, doodle, come and get some bread and butter;
Doodle, doodle, come and get a barrel of sugar.
Doodlebug, doodlebug, come and get a grain of corn.
Your house is burning up.
Doodlebug, doodlebug,
  Come out of your hole;
Your house on fire
  And your children will burn.
Doodlebug, doodlebug, come out of your house;
It's burning up with your wife and all of your children,
Except Mary – she's under the dishpan.
Doodlebug, doodlebug,
  Come out of your hole;
If you don't,
  I'll beat you black as a mole.
Doodle, doodle, doodle,
Your mother and grand-daddy are dead.

(Hand 1964)

While the origin of charms like this are unknown, it is speculated that they stem from European culture. It also is speculated that the violent themes presented could have served as reminders for children to obey their parents or they will burn up and die.

The history of European relations with Native Americans is a history saturated with murders, looted graves, illegal land transfers, and disruptions of sovereignty. In the seventeenth century, Puritan writings described Native Americans as demonic manifestations of a psychic struggle. During the Enlightenment, European American writings invoked natives as symbols of internal darkness and irrationality, when in fact it was the colonists who had plagued them— metaphorically and literally (Bergland 2000).

Many legends about lepidopteran species exist in Native American lore. Due to their beauty and extensive metamorphosis they possess the power to facilitate a deep reflection on oneself. The Nizitapi (Blackfoot) held a belief that dreams were brought to us by butterflies. Women often embroidered butterflies on hair ties for babies and children to bring them good dreams. The Hopi people also have belief systems revolving around butterflies; for example, the butterfly dance (Cherry 2019), which is a petition for rain, good health, and longevity. It also gives thanks to the butterfly for its beauty and contribution to pollination. They dance in a two-day event and only young, unmarried girls are allowed to participate. The girl's partner during the ceremony must be from her father's clan and they must abide by the rules of the ceremony, which lay out what is considered to be appropriate behavior. The girl must learn thirty-two songs that will be performed during the two days and the family dresses the girl for the dance in ceremonial clothing. The village plaza will be prepared, and after the preparations the dance will begin. During the dance the girl will wear her kopatsoki in a humble way and after the ceremony she will receive it as a gift from her partner along with a basket of groceries. Over the next few weeks her family will repay the boy's family with traditional Hopi foods. This exchange of gifts is a way of recognizing and accepting special family relationships (Lomahaftewa n.d.). The spirit of the butterfly is represented during these sacred dances (Cherry 2019).

The butterfly also represents rebirth, regeneration, and joy for Aztec people of Mexico. One legend states that Quetzalcoatl first encounters the world in the shape of the chrysalis. The deity Itzpapalotl then emerges in a burst of beautiful painful white light. She is the patron deity in Aztec culture, and it is said that dead family members visit those still walking the Earth as butterflies. Numerous other examples of lepidopteran mythology among Indigenous peoples exist and provide individuals with the opportunity to observe the manifestation of growth and regeneration (Cherry 2019).

The Goajiro people in Colombia regarded a large white moth in a bedroom as a visiting ancestor's spirit and took great care in their presence. The Bolivian group Aymara regarded a rare type of moth's visitation as a death omen (Cherry 2019). In *El Orinoco Ilustrado*, Father Joseph Gumilla, S.J., as well as others, documented near the Orinoco River (what is now Colombia and Venezuala) an intriguing association between a bug called a *coya* and a bonfire lit from grass called guayacán. Gumilla was a missionary in this region from 1715 to 1750 and provided information about the people as well as the flora and fauna of this area in his book in 1741. He described a coya as bright red and tick-like. If one is accidentally smashed on a person's skin, the body fluids of the crushed organism will cause swelling on the person's entire body, eventually killing him or her. The only way to prevent this demise is by allowing four or five men to pass the victim's body over the flames of a bonfire built from guayacán grass, scorching the entire body.

Gumilla stated that other animals, such as cows, also are weary around coyas and avoid trampling them. The insect is not reported to bite or sting, but only causes harm when crushed on skin. Other accounts refer to the coya, or coyba, as a spider with a red abdomen and claim ingestion of human excrement could be a cure, in addition to the bonfire, as well as a newer, Christian remedy of lemon juice boiled with sugar. Others claim that oxen and cattle are safe from harm because the coyas hide when they detect the heavy footing of these animals and therefore would not be trampled (Kamen-Kaye 1979).

Colonies during the seventeenth century had a culture that included pieces of cultures from England, Africa, and Dutch and other English colonies, and people still maintained ideas about insects that were passed along as folklore and old wives' tales. A spider's web baked into an apple was said to cure chills and fever, but according to Swedish naturalist Peter Kalm, who visited the colonies, if three apples were eaten, the ailed person may fall into a strong coma, but may wake later on and feel fine. Colonists mainly documented their experiences with insects as pests, particularly problems with mosquitoes in Plymouth and Jamestown. Other documents relate issues with cockroaches and flies, not to mention lice and weevils tormenting the pilgrims and their food on the sea voyages from Europe to America. English doctor Thomas Mouffet also wrote about cockroaches in 1658 and showed great interest in insects, which made him the origin of the nursery rhyme "Little Miss Muffet," in which a spider sits next to a girl, frightening her away, even though many spiders in the region were not dangerous to humans. Another interesting colonial appearance with insects is in the making of the famous red coats worn by British soldiers that were dyed with the pigment from cochineal larvae in the order Hemiptera, originating from South America (Robinson 2007).

The relationship between China and North America during the Age of Enlightenment has been largely ignored and many historians have assumed that the Chinese culture did not penetrate North America until the establishment of trade between Canton (known now as Guangzhou) and the East Coast after the War of Independence, but the Chinese had a much larger influence during the American Enlightenment than just trade. For example, in 1733 James Logan, a prominent wealthy friend of Benjamin Franklin, had acquired several books on Confucian philosophy (Alridge 1993). While many Chinese beliefs developed long before the Enlightenment, these beliefs made their way into the New World through books and travel during this time period. Social insects, in particular, offer a source of allegorical meaning for human life (Coleman 2006). For example, the written symbol for ant is a combination of *righteous* and *insect* which references the subordination and orderly marching of ants. Ant nests are perceived as being perfectly built and organized with compartments for previsions and areas to nurture their young, and Chinese writers depicted the systems with impeccable accuracy. The ant is the embodiment of virtue and devotion, but also symbolized self-interest and filthy profit, which led to the development of the Chinese expression "ants cling to what is rank smelling" (Williams 1976).

The cicada is another common insect in China that provided allegorical meaning to human life. The cicada belongs to the order Hemiptera, which emits a sound caused by the production of vibrations with the lamellae, which are two flaps that come together causing a unique type of music. A cicada nymph will spend many years of its life underground, eventually finding its way to the surface, then molting to reveal the winged insect. This life cycle is an emblem of immortality and resurrection in ancient Chinese mythology. This belief was exercised by placing a piece of jade carved into the shape of cicada into the mouth of the recently deceased prior to burial (Williams 1976).

Entomology was more than just the study of insects during the Age of Enlightenment. It allowed society to mourn the dead and sing songs about their lost loved ones, and insects facilitated the disempowerment of the Catholic Church, which inevitably led to the Protestant Reformation and large migration to the New World. Exposures to new cultures through European expansion produced a whole new culture and a blending of old Native traditions with new European ones. Society knows that we need insects from a health and pollination perspective, but it is rare that we reflect on the societal and cultural aspects of entomology. Religion, mythology, folklore, and oral traditions have shaped the entire world, which was facilitated by insects.

CHAPTER SIX

Insects as Symbols

JAMES N. HOGUE

Human thought and understanding are often thought of in terms of light, or a shedding of light that exposes the reality of the nature of the cosmos. This is what is implied with the Age of Enlightenment (1600–1820). This time period is defined by a set of attitudes and practices associated with rational acquisition of knowledge; understanding reality through experience, observation, and experiment. Books such as the well-read "*The discoverie of witchcraft*" by Scot (1584) exposed popular, and even scholarly, belief in witchcraft, magic, and other superstitions as irrational and un-Christian. The preceding age of the Renaissance and the Reformation reinforced the authority of sacred texts and European education was based on a revival of Greek and Latin sources. Knowledge was therefore based on two authoritative sources: scripture and the classics of the Greco-Roman world (Hampson 1968). The Enlightenment was a time of change in the intellectual environment in much of Europe and in the American colonies that would become the United States of America near the end of this era. On the other hand, the rest of the world experienced little change associated with the Enlightenment and remained under the grip of long-held ideals of the past.

To many, this was also a time of anticlericalism, sometimes tending to antireligion, where this exercise of reason fulfilled a newfound desire for the systematic and scientific investigation of nature that would reveal the natural and moral laws by which nature operates rather than validate perceived religious norms. So, to some, this rationalism was very disruptive as it was replacing earlier and sentimentally comfortable ways of knowing. Consequently, this new and different way of understanding in the history of human endeavor was controversial and led to discontent that became manifest in society through various means, including art and literature. It is through these and related avenues of discourse that insects and related arthropods get co-opted to symbolize aspects of both enlightenment and counter-enlightenment ideology. Insect symbolism during this time is thus partly a product of the times.

The discontent being voiced was from a populace that liked doing things the old way. This has been true throughout human history as change inevitably takes place. The old ways are stubbornly persistent and are not supplanted easily.

A reasonable expectation in hindsight is that at least the members of a civilized and somewhat educated society would leave behind any world views that have been revealed as myths, superstitions, or falsehoods by modern science and embrace new, empirically generated knowledge as truth and behave accordingly. But this is not how people behave. For example, the long history of the development of evolutionary theory is marked not by the quick abandonment of falsified ideas and the incorporation of well-supported hypotheses into current ideology, but by the clinging to unfounded notions of the past, some of which trace their origins to the ancient philosophers of Greece. Such notions served as impediments to the progress of understanding and the development of ideas upon which new hypotheses are generated and tested (Mayr 1982). Even today, a large part of the population continues to hold views of the origins and workings of the natural world that are counter to the current scientific understanding of these issues (Davidson et al. 2017). New kinds of knowledge and ways of knowing were supposed to supersede supernatural means and explanations at this time and yet change was slow. Therefore, it is not surprising that in an era that what would otherwise be dominated by reason, objectivity, and truth, folklore and myth continue to exist in the populace. For example, during the Age of Enlightenment, many people continued to believe in witches, ghosts, fairies, and the like, and therefore continued to incorporate them into varied aspects of their lives (Thomas 1971). Museums and other public and private displays of natural history objects at this time were also sites that stubbornly perpetuated misinformation and supernatural properties of organisms, even in the face of well-founded knowledge to the contrary, in order to maximize the mystery and wonder of their exhibits (Impey and MacGregor 1985).

Myths are not only inherently symbolic in and of themselves (Kirk 1970; Kritsky and Cherry 2000), but they additionally incorporate specific instances of object symbolism. The latter is also true of folk tales. These symbols are derived from a wide variety of man-made and natural objects, including many plants and animals. Because insects and other terrestrial arthropods are very apparent and biologically influential elements of the human experience, it is not surprising that they have become incorporated into symbolism. In these cases, insect symbolism employs the image or verbal description of insects to signify some idea other than the insect itself (Hogue 1975). This is the other means by which insects were employed symbolically at this time. They are carried over in both practical and irrational applications from an earlier time rather than being a product of the times.

Therefore, people continued to consciously employ insects symbolically during the Age of Enlightenment for a variety of reasons. Insect imagery was used as a tool to make a point or to educate the populace in current, often contentious, matters of science, art, religion, and politics. Insect symbolism also existed in its more ancient form as a carryover from earlier times in a variety of contexts, including mythology and folklore. Lastly, insect symbolism is found as part of human entertainment, such as in storytelling, simply because it is fun. And of course, the symbolism of insects never left our subconscious, even in this enlightened age, and thus their images and actions continued to both enrich and haunt our dreams (Klein 2012). The human mind of the enlightenment and its associated psyche still retained its own psychological and sociological history, inherited from our cultural and evolutionary past. The following examples will serve to illustrate how insects and their relatives were used during this age as symbols of human ideology, myth, and entertainment and that in turn influenced related human endeavors in subsequent times.

LITERATURE

The primary avenue through which popular discontent was portrayed between enlightened thinking and practices of the past was in literature. A good example involves the "Quarrel of the ancients and the moderns," constituting recurring instances of antagonism between ideology of the past and modern, progressive ways. This played out in seventeenth-century France and England where prominent humanists argued that recent scientific and technological advances demonstrated an independence and superiority of modern culture compared with that of the classical authorities of Greek- and Roman-based schools of thought, which argued that current advances were based on a foundation created by Greek and Roman ideology. Contemporary writers were accused of simply imitating the ancients. Although this was primarily a literary and artistic debate played out in literary circles, these arguments had undertones questioning the authority of long-held political forces and religion and the relevance of the voice of the people (DeJean 1997). This contentious discourse played out in the literature of the times, and in some cases, this involved the symbolic use of insects to convey points of view. One such work was Jonathan Swift's *The Battle of the Books* (Swift 2010). In this short work (written in 1697), Swift describes a literal battle between the ancients and the moderns that takes place in the library of Saint James as enacted by animated books representing ancient philosophers and modern idealists and critics. This battle is interrupted by an allegorical digression describing an argument between a spider in its web in a high corner of the library above the shelves (Figure 6.1) and a bee that has wandered into the library through a broken window and ends up destroying part of the spider's web. The angered spider launches into disparaging the bee for being a homeless vagabond, without stock or inheritance, a freeloading plunderer of nature. The spider then extols his personal virtues of being a domesticated creature that builds his own home with his own materials. The bee counters by describing his work ethic, pointing out that he takes from but doesn't injure nature, and furnishes mankind with two wonderful things, sweetness and light. Although the argument between the spider and the bee includes talk of respect and tolerance of others and is conducted with civility, it is clearly a parallel play on the much less civil ideological battle going on below among the books.

FIGURE 6.1 *The Battle of the Books*. Note the spider in its web and bee at the top of the window. Wikimedia Commons.

While the impartial author does not indicate a winner among the battling books, the interactions of the spider and bee, through dialogue and symbolic association of known biological traits, illustrate the disagreement in points of view and self-image. The spider, with his complicated yet fragile web, represents the moderns and sees himself as independent and displaying great skill in architecture and mathematics in the design of his webbed fortress. The bee represents the ancients, disparaging the accomplishments of the moderns as possessing little merit and no durability; living off of the accomplishments of others, sucking them dry like the multitude of exsanguinated fly carcasses that are described lying about the spider's domain. The bee in turn creates good things through cooperation and collective wisdom. His merits come from heaven alone and not others.

Literary insect symbolism is also found in poetry and can serve a similar purpose or be used to ponder meaningful questions about life. Such is the case for William Blake's 1794 poem "The Fly."

Little Fly
Thy summers play,
My thoughtless hand
Has brush'd away.

Am not I
A fly like thee?
Or art not thou
A man like me?

For I dance
And drink & sing;
Till some blind hand
Shall brush my wing.

If thought is life
And strength & breath;
And the want
Of thought is death;

Then am I
A happy fly,
If I live,
Or if I die.

(Blake 1794)

The actors in this poem, originally printed as part of a painting with the text set within the branches of two trees that frame a woman playing with two children, are a man, a fly, and the man's thoughts of life and death. The fly here symbolizes a natural insignificance, that is imparted on the human poet speaking the lines as part of his contemplation of the meaning of life and his role here on Earth. The poet brushes away the fly; so might not he also be brushed away as a small bothersome and trivial creature by God? The fly does not focus on such sorrow, so rather than being scared of death, we should focus on living life.

Bernard Mandeville (1924), in the explanation of his 1714 poem "The grumbling hive: or, knaves turn'd honest," uses the hive (of honey bees) as a symbol of a successful nation. He proposes that society functions only because there is a measure of poor behavior among the citizenry. Society cannot generate and enjoy the comforts of life that come from a wealthy, industrious, and powerful nation and simultaneously be home to a virtuous and innocent populace. The roots of group prosperity and a flourishing society are the individual sins of the people. Private virtue, on the other hand, leads to economic depression and societal ruin; hence the phrase "private vices—public benefits" in the title of his principal work on the subject (Mandeville 1924). This comparison between hives of eusocial insects, especially those of familiar honey bees, is an ancient and commonly employed analogy (Farrell 1985). Mandeville is pointing out corruption and hypocrisy in English society and government, but he is also apparently disparaging the beehive–human society comparison as inaccurate and ineffectual. Bee colonies are not models for humans, but insects that merely behave in part similar to people (Farrell 1985). Of interest is that Mandeville's description of the division of labor, stemming from one of the attributes of eusocial insect colonies, was taken and developed later (1776) by the Scottish economist Adam Smith in his famous work "An inquiry into the nature and causes of the wealth of nations."

Other works of the times, written in a poetical format, are less serious and not as philosophically deep. Several writers used insects, often in anthropomorphized situations, as symbols of whimsical fun that serves as amusement for the reader. Such fantasy writing is most often directed at children, who might see the world of insects as happy and enchanted. In William Roscoe's poem (1808) "The Butterfly's Ball and the Grasshopper's Feast," a host of insects and other small animals gather for a party put on by a butterfly. The characters set aside their everyday tasks to spend time together and have a good time, in a world all their own, and yet a world somehow familiar and relatable to the minds of children who have peered at the lives of these little creatures in the fields and woods near their homes.

The antagonism towards change and longing for the old ways also played out in new genres of literature. Books written for children developed significantly in the 1700s and many featured stories of other worlds inhabited by a host of creatures from regional folklore and their interactions with humankind. In the broad sense, such "fairy tales," enhanced by creative and fanciful illustrations, linked children (and adults) to an age that was disappearing in the face of scientific inquiry and yet familiar and longed for. Thus, literature and other forms of imagery served to counter the onslaught of science and reason upon the familiar and comfortable ideals of the people. One response was the depiction of fairies and other supernatural beings in works of literature and art.

FAIRIES AND THEIR WINGS

People worldwide have incorporated fanciful creatures into their religions, mythology, and folklore. Sea-faring people have their sea monsters, mermen, and mermaids and people of the countryside and woods have their forest spirits, both good and bad. Such is the case with the fairies and related host of woodland nymphs and sprites; and nowhere did this develop to the extent it did in the British Isles.

The fairies of England, Scotland, and Ireland likely have their origin in mainland elves and dwarfs, particularly those of the ancients of the Scandinavian Peninsula (Keightley 1968). A Cornish belief is that fairies were the souls of ancient people too good for Hell, but too bad for Heaven, and that they had become ethereal versions of their prior selves. Others were equated with souls of the dead, versions of the deceased that leave the body at death, or later as fallen angels. Some fairies were thought of as beings that are between material creatures and spiritual entities that are typically invisible, living in an invisible world (Briggs 1976), in some cases in a world between heaven and hell (Latham 1930). In most cases, the origins of fairies are not very clear, but the fairies of the Middle Ages were certainly not small nor were they particularly kind (Thomas 1971), often being associated with witchcraft (Latham 1930). These legendary versions of fairies were typically thought of as just smaller in size than a typical human, or the size of a child, and were wingless (Latham 1930). There were some smaller fairies, and like their larger counterparts, these were also wingless and would move about or fly using magic to levitate themselves or they would ride bits of plants (stems, twigs, grass), sort of like a witch riding a broomstick (Briggs 1976). Others rode upon birds or other creatures, including insects. Such were the notions of fairies in England until the time of Shakespeare, when they transformed into a smaller race of mischievous, but generally friendly, creatures.

The fairies in Shakespeare's *A Midsummer-night's Dream*, written in the mid-1590s (Shakespeare 1908) were of this kind. This successful play, and later adaptions of this story such as Purcell's (1692) semi-operatic "The

Fairy-Queen," were performed during much of the seventeenth and eighteenth centuries and popularized this different version of fairy. Some of the fairies in this story (e.g., Oberon and Titania, king and queen of the fairies) are of traditional English fairy stature, of nearly normal human size, whereas others (the attendant courtiers of the royal pair) are much smaller (Schanzer 1955). Furthermore, these fairies are dream-like, beautiful, and they belong to the enchanted world of the winds and nature (Schanzer 1955). These fairies also dance (Latham 1930). By the beginning of the 1600s, fairies are regularly being depicted as tiny creatures. For example, the forest fairies in the stage play *The Maydes Metamorphosis* (Anonymous 1600) are small enough to sit upon flowers or to ride around on flies. They are also fond of singing and dancing, a favorite behavior of this new type of fairy of this era.

The popularity of this newer version of fairies and the contemplation of their idyllic fairyland world was an escapist fantasy by people of this era fascinated by the supernatural and romanticism that was fostered as a response to a widespread rejection of the values during the Age of Enlightenment. These new fairies are familiar and likeable. They were local products, not from a foreign land, not spiritual manifestations of the dead (and so not scary), and they were not as powerful as gods, only performing minor, primarily mischievous, acts of magic (and so not to be feared) (Bown 2001). They are of a world inhabited and animated by the supernatural. Similar to pixies, they are often implicated in the behavior of misleading travelers. They can influence the behavior of humans and otherwise cause mischief among those people that spark their displeasure. These fairies are as real as dreams, the dreams of play, lightness, and a magically beautiful vision of nature that gave a sentimental feeling to people in what they viewed as a time of deadening rationalism in the modern forms of thought, technology, and social organization (Bown 2001). It was as if the Enlightenment had destroyed the world and culture that fostered the fairies in the first place; the world of the common folk and their traditions. Fairies are from an unenlightened age, a time of the common folk, before knowledge of true religion and before the truth of the world was revealed by science (Bown 2001). A world longed for by the people.

The most notable characteristic about fairies of this era is that they have wings. The wingedness of Shakespeare's fairies is not much elaborated until later times in other works by illustrators, such as Henry Selous (Clarke and Clarke 1886), Arthur Rackham (Shakespeare 1908), and Walter Paget (Lamb and Lamb 1910). These flighty characters evoke a spiritual, ethereal, and dream-like mood to the story. This impression is enhanced by the imagery of otherworldly beings bearing the wings of spirit-world counterparts of real-world creatures. Also, the kind of wings they bear means something; they are symbolically significant. Typically, angels bear bird wings, Satan and evil dragons have the wings of bats, and fairies sport a host of insect wings, particularly those of butterflies.

Early in the eighteenth century, Alexander Pope described insect wings on his fairies (or sylphs) found in his poem "The Rape of the Lock" (1717). Later, Thomas Stothard represented these fairies bearing butterfly wings in his 1797 illustrations for Pope's poem (Figure 6.2). The choice of butterfly wings was apparently suggested by William Blake, whereupon Stothard went out and collected butterflies upon which to base his fairy wings (Brittan 2011).

Angels have wings, not really to fly, but as a symbol of their spiritual nature and their ability to move between heaven and Earth. Angels come down from heaven, fairies come up from the Earth. Likewise, fairies don't need wings to fly, they are symbolic of their abilities and qualities. These wings signify spirit,

FIGURE 6.2 The origin of insect-winged fairies. Thomas Stothard's engraving depicting Pope's winged fairies from the 1797 edition of his poem "The Rape of the Lock."

power, and swiftness (Bown 2001) and they are of the Earth. The insect-based wings of modern fairies also link these little inhabitants of enchanted fairy land to the real, but no less mysterious world of insects.

The real world of insects is a miniature world, much like the diminutive, yet fantastical world of fairies. The contemporary interest in the tiny world of fairies and the newfound knowledge of the similarly minuscule lives of insects, as revealed by novel breakthroughs in microscopy, led to a perceived equivalence of these domains (Brittan 2011; Forsberg 2015). This is a convergence of nostalgic folklore with a burgeoning field of science. Scientists using microscopes used the fantastic to capture the imaginative power of what they saw and wanted to share, and writers about fairies incorporated a scientific vision of this world to depict a sense of reality (Brittan 2011). The insect-based wings on fairies symbolize a sense of reality because they are associated with flight and they link fairies with the similarly miniature, yet real, world of insects. In addition, these ever-active, dancing fairies are often associated with plants, especially their flowers; yet another link to real insects.

Sweet flora as if ravisht with their sight,
In emulation made all lillies white
For as I oft have heard the wood-nimps say,
The dancing Fairies when they left to play
Then blacke did pull them, and in holes of trees
Stole the sweet honey from the painfull Bees,
Which in the flowre to put they oft were seene,
And for a banquet brought it to their Queene.

(Browne 1845, written around 1616)

Again, the progress of science precipitated alienation of the common people and an ideological rebellion against this inflexible, dehumanizing rationalism of the day. Depictions of the irrational served to exemplify this rebellion to a sensitive audience.

In general, insects symbolize another world that is also available for examination and study and as the focus of our wonder. Furthermore, it is relatable through the analogy with the minuscule fairy land and the similar bizarre beauty of insect and fairy forms and behaviors. Such depictions of the fairy world would reach impressive development not long after the Enlightenment in innumerable examples throughout the literary and visual arts. The fascination with fairies significantly bloomed in the late 1700s and existed for over a hundred years, reaching its peak later during the Victorian era. A fine example of the equitable interactions of the fairy and insect worlds is Charlotte Maria Tucker's, who wrote under the pseudonym A. L. O. E. (a lady of England), *Frisket Fairy* (A. L. O. E. 1874). In this delightful set of stories, a diminutive, winged fairy (Frisket) introduces the reader to a variety of insects and their habits through narrative and drawings. Victorians were the connoisseurs of the fairy (Bown 2001).

This genre of fantasy also would soon lead to a different kind of melding between the human and insect world in the wonderful illustrations and verbal descriptions of anthropomorphized insects (or perhaps ento-morphized humans) by artists, such as Martin Disteli, J.J. Grandville, and L.M. Budgen (Schachat 2014). Grandville (Jean Ignace Isidore Gérard of early to mid-1800s France) often employed his creations for political caricaturization and to criticize various aspects of European society (e.g., see Hetzel 1977). Once again, a reflection of the popular discontent the people had with the current social order and its rational arguments against the sentimental magical view of nature. Budgen's (writing under the pseudonym Acheta Domestica, 1851) text and illustrations were used primarily to entertain and instruct the reader about the wondrous goings-on of insect lives. Her humanized insects were sometimes likened to fairies or possessing fairy-like qualities; qualities she even associates with the endeavor of entomology itself.

Although beyond the temporal scope of this article, these works precipitated on the one hand a proliferation of a wide variety of insect-inspired characters, that is "bugfolk" (Hogue 1979), within a host of literary and illustrative media. And on the other hand, these works also are relevant in that they paved the way for a more scientific study, description, and illustration of insects. One of the earliest general works on insects is Thomas Muffet's (1634) *Theater of Insects* that deals primarily with the classification, characteristics, and natural history of insects. But his discussion of insect biology also incorporated much speculation, folklore, poetry, and superstition. He also employs his subjects to symbolically portray both positive and negative aspects of life based on their characteristics described therein. For example, flies are honest creatures that delight in the truth because they go about their activities only in the light of day, considering it unbecoming to be out at night. The text contains numerous poems and is rife with speculation and folklore about the behavior of insects and thus exemplifies a bridge between insect fantasy literature and scientific discourse. Later works, even after the Enlightenment, remained true to this form (e.g., Cowan 1865), but provided a well-received introduction to the insect world and became the precursors of scholarly works describing real insect biology. It is here that we have the beginnings of the parting of ways of science and fantasy. A phenomenon that owes part of its history to woodland fairies and their insect-inspired, symbolically charged wings.

ART

Literature and its associated illustrations provided a prominent venue for the depiction of insects in symbolically relevant ways. Art without text provided a similar opportunity for the artist to use insects to give voice to an opinion, provide additional meaning to a painting, or simply to show off. For example, many paintings of the Renaissance era used flies as nature-inspired motifs for a variety of purposes, including what is known as *trompe-l'oeil* (optical illusion), a technique with ancient roots whose purpose is to visually trick the eye of the viewer. For example, a fly might be painted at a different scale from the subject of the painting and positioned as to appear to be sitting on the frame. In other situations, insects such as flies were included in paintings to demonstrate the technical prowess of the artist to clearly paint something so texturally complicated as an insect (Chastel 1984, 1986; Connor 2007).

In the era of European exploration and colonization that preceded the Age of Enlightenment, some European countries that were involved in maritime exploration and trade became extremely wealthy. Resident artists often included objects, particularly food items and biological curiosities, that reflected a particular country's access to products from far-away lands and a blossoming global market economy. Such paintings showing food were thus not realistic depictions of what the people ate, but rather served to advertise the artist's, or patron of the artist's, aspirations and aesthetics. Painting such things in this context that were hard to come by was a reflection of a nation's or family's wealth and extravagance and an impression of their worldliness (Wansink et al. 2016).

One place where this was particularly realized was in the Netherlands, a place that became the wealthiest nation in the western world in the second half of the seventeenth century (Bryson 1990). During this Dutch Golden Age (1585–1735), the Dutch masters of still life paintings employed symbolism through objects depicted in paintings to show off their indulgences as everyday affairs and what they as a society had access to. In addition to food items, seemingly odd things that were apparently inconsequential to the main subject were included in a manner that is sometimes called "disguised symbolism" (Panofsky 1966). Such things included seashells and insects from far-away places, again symbolizing the nation's cosmopolitan opulence. One example, located in the Toledo Museum of Art, is one of several paintings by Jan Davidsz de Heem (1606–1683/4) referred to as "still life with a lobster" (late 1640s). This painting depicts a table overflowing with fruit, flowers, tableware, a lobster, and no less than eight insects (three butterflies and single representatives of a bumblebee, blister beetle, lepturine longhorned beetle, a caterpillar, and what appears to be a leptogastrine robber fly). See also Bergström (1983); especially the exquisitely rendered insects in the works of Jacques DeGheyn, Balthasar van der Ast, and Ambrosius Bosschaert the Elder.

Another, symbolically similar and primarily artistic, use of insects was their collection and display in decorative arrangements of "naturalia." Collection of natural history objects was a recognized pastime of the wealthy, and to some extent of middle-class folks in many parts of Europe at this time. Specimens of precious stones, shells, coins, and even vertebrates were gathered and displayed in artistic fashion in cases. Insects were featured prominently in these displays. Those from far-away lands were especially prized and, like their counterparts in still life paintings, they served as symbols that advertised the worldly status of the owner and of the nation (van de Roemer 2004). These cabinets of rarities (Dutch *Rariteitenkabinet*) or natural and artistic wonders (German *Wunderkammern, Kunstkammern*) were the predecessors of modern-day natural history museums (Impey and MacGregor 1985; Neri 2011).

Christian art of this era also used insects symbolically to add meaning (Sill 1975). In this context, bees are a symbol of hope for the soul. They work tirelessly and their appearance each spring symbolizes rebirth. This rebirth led bees to symbolize Christ. Honey bees possess purity and sweetness (their honey) and yet their sting serves as a reminder of Christ as judge (Werness 2004). Bees also are a symbol of industry, creativity, and wealth because of their communal organization and ability to make and store honey (Sill 1975). An old and mistaken belief that bees were parthenogenetic (could reproduce without being fertilized by males), or at least recognition that the workers do not have sex, resulted in bees symbolizing the Virgin Mary and the Immaculate Conception. Flies represent a bearer of bad tidings, pestilence, evil, and are often a symbol of sin. In Madonna and child paintings, flies represent the evil that Christ is to experience and overcome (Impelluso 2004). They also depict death and the vagaries of life because of their real-life association with destruction and putrefaction. Locusts, or grasshoppers in general, are forces of destruction in reference to the plague of locusts that swarmed and decimated lands in Egypt because the Pharaoh would not submit to the wishes of Moses. Spiders are also synonymous with evil and Satan. Because real spiders catch unsuspecting flies in their webs, their behavior is analogous to the deceptions of Satan. Spiders also represent miserly behavior in people who suck the life out of those indebted to them. Lastly, spider webs, because they are so fragile and seem to be spun out of thin air, are likened to the fragility and weakness of human existence (Sill 1975).

The symbolic use of insects in art was probably most developed in what is known as "Vanitas" paintings. This descriptor comes from the Latin word "*vanus*" for emptiness and refers to a view (based on one Christian understanding of our earthly existence) that our mortal lives are without much merit, and thus empty and futile, compared with our life in the hereafter. These paintings also illustrate the problem of affluence and advocate a denial of material excess, in a somewhat hypocritical way given the substantial worth of these paintings (Bryson 1990), and that life is fleeting and death is always near. Objects in these paintings thus directly (e.g., skulls) or indirectly (e.g., flies) use imagery depicting death and also symbolically incorporate objects symbolizing the transience of life (e.g., insects such as butterflies, hourglasses and candles, cut flowers). One relevant example is that of a north German painter from around the year 1600 (Figure 6.3). The skull and femur are clear images denoting death, whereas the frog, that lives in swamps, is a less direct allusion to our ultimate fate. The cut

FIGURE 6.3 Vanitas painting by a North German painter *c.* 1600. Source: Impelluso 2004, but from Dortmund Museum fur Kunst und Kulturgeschichte.

flowers, very common elements in vanitas paintings, indicate the transience of life and beauty. Just when they reach the height of their visual appeal, they begin to wilt and fall apart. Readily associated with death and decomposition, the fly on the skull is another clear symbol of the passage of life into death. Insects in general are used in this context as symbols of our mortality and that time is short and subject to deterioration (Impelluso 2004). This stems from the general perception of many insects as agents of decay. Lastly, the Latin phrase in the image is from the Old Testament passage Job 14:2: "We blossom like a flower and then wither, like a passing shadow" (New Living Translation). Together these images drive home the point that human life, much like the short lives of insects, is fleeting and of little consequence.

The detail and realism in some of the richly adorned earlier works depicting insects, such as the marginal illuminations and vanitas paintings (e.g., *Allegory of Life and Death* [1598]) by the Flemish artist Georg Hoefnagel (1542–1601) not only served as inspiration for works during the Enlightenment, but paved the way for scientific illustration of later periods.

HERALDRY

Not all of the symbolic uses of insects took place within the context of societal disillusionment with changes brought about by the Enlightenment or as ways to advertise the wealth and prosperity of a people. Some were carryovers from previous ages that continued to find relevance in people's lives at this time or from our more distant past that are difficult to abandon. For example, images of insects were featured on hundreds of ancient Greek, Roman, and Ephesian coins and some of these images conveyed meaning through symbolism (Brown 1993). Although to a much lesser extent, some coinage, medallions, and tokens continued to feature insect imagery during the Age of Enlightenment. Such was the case for heraldry.

From the earliest of times, humans have used symbols to identify themselves, their families, their clans, and their armies. The systematic means by which this is done, both on a nationalistic and familial level, is known as heraldry. Although heraldry is a phenomenon primarily of western Europe and practiced by "gentlemen," it also existed, but perhaps to a lesser degree, in America (Allcock 1962), and some other parts of the world, at least in the broadest sense of the concept. Among the earliest arms illustrated are those dating from the mid-1200s in France (Dennys 1982). The fourteenth century was the golden age of chivalry (with heavy incorporation of heraldry), but heraldry remained active throughout much of the Age of Enlightenment then fell out of favor near the end of this era, around the time of the French Revolution at the end of the eighteenth century (Allcock 1962). A similar disfavor happened later in

America as arms were considered by some to be undemocratic and snobbish (Allcock 1962).

People of semi-civilized nations have a long history of assuming the virtues and the names of animals and deities and using them as symbols of honor (Fox-Davies 1968). The heraldic achievement (commonly known as a coat of arms) consists of many parts, of which the centrally located shield is among the most important (Dennys 1982) and is the element typically bearing the arms (designs) or charges that may be representations of living or inanimate objects with symbolic significance. This is where insects are typically found. Insects on arms include butterflies and moths, bees, flies, wasps, cicadas, beetles, grasshoppers and crickets, and fleas. There are even a few with scorpions and isopods. Bees are by far the most common design and they are usually depicted flying (volant) and in dorsal view (Keiser 1966; Cussans 1970). For example, the shield on the coat of arms of the Rowe family features a beehive surrounded by flying bees (Cussans 1970) and a grasshopper sits on the crest of the Gresham family arms. Bees represent community and industry and grasshoppers were esteemed by the ancients as possessing wisdom (Porny 1795); crests bearing these insects suggest these attributes on the associated human families. Fox-Davies (1968) notes that bees feature most prominently among the insects in European heraldry, being found as charges on the shield and crest. He notes that the Emperor Napoleon chose a bee to adorn his heraldic badge and arms, which in turn gave bees increased standing within French heraldry.

Other insects mentioned by Fox-Davies are ants, butterflies, flies, silk moths, and stag beetles. A good example of stag beetle charges is found in the arms of the Schröter (see also Schroeder, Schrader, Schrodinger, and several other spellings) family (Figure 6.4; see also the coat of arms of Johan Wilhelm Schröter, 1699, found in Klausnitzer 1981). Schroeder or Schröder is an occupational German surname for a tailor, derived from the Middle High German word "schraden" or "schroten" (to cut). The connection to the beetle is likely through the larval feeding behavior of gnawing, or cutting, through decomposing wood. German people of the time called this insect Schröter (Adams 1859; Klausnitzer 1981), meaning "nibbler" or "nipper." The English referred to it as "nippy-dor" or "pincher-bob" (Adams 1859). A close relative in eastern North America, *Lucanus elephus*, is similarly referred to as "pinch bug" in reference to its ability to deliver a strong pinch with its mandibles. There was added benefit from association with stag beetles as they were symbols of Christ, because of their equivalence with male deer (stags) that have been holy animals since classical antiquity (Werness 2004).

In addition to familial arms and those attained during their career, the Pope and the Papacy had their own set of arms. Relevant to the Enlightenment, the Arms of Pope Urban VIII (1623–1644) (born Maffeo Barbarini) are charged with his family symbol of bees. These bees soon came to decorate paintings,

FIGURE 6.4 Coat of arms bearing charges of the male European stag beetle *Lucanus cervus* (Coleoptera: Lucanidae). Drawing by J.N. Hogue from Siebmacher 1906.

sculptures, and buildings (including St. Peter's Basilica and Urban VIII's tomb there) of the Vatican. At one point, some ten thousand painted or carved examples of bees existed here, primarily the result of the unfettered exuberance of the Pope's close friend and architect Gianlorenzo Bernini. The Barberini family was a noble and influential family in Rome. Their earlier family name was Tafani, or Tafani de Barbarino, and their arms were adorned with three horse flies. Tafani is Italian for horse flies. As a means of increasing the family's social status, Maffeo did away with the negative symbolic connotations of flies by changing the family name to Barberini and changing the flies to bees. In addition to the bees, he also added the sun and laurel associated with Apollo to the family emblems, thereby cultivating his desired poetic image and aligning himself with symbols of hard work and dedication (Majanlahti 2006). These new images were also reinterpreted in Christian terms: the sun referred to wisdom that was Christ, the laurel represented eternal life, and the bees were seen as wise and chaste, also attributes of Christ (Majanlahti 2006). Interestingly, in 1630 Francesco Stelluti produced a book of poems dedicated to the Cardinal Francesco Barberini, the nephew of the current Pope, Pope Urban VIII. The Barberini family crest included a trinity of volant bees, so Stelluti incorporated this emblem onto the frontispiece of this work that included his 1625 plate of honey bee drawings. These illustrations are apparently the first drawings in a printed book of an organism made with the aid of a microscope (Ford 1993).

The notion of bees (specifically honey bees) as creatures worthy of adoration followed them across the Atlantic with their introduction into the New World in the early 1620s. They served as symbols of industry, community, and the benefits of hard work to many settling new lands in North America. For example, honey bees and the skep style of beehive (a bell-shaped structure constructed of coiled straw, wicker, or rope) were adopted as emblems touting these virtues in the late 1840s by Mormon settlers in what would soon become the Territory of Utah. A skep is currently part of the official seal of the state of Utah.

Heraldry and the associated symbolism of design is not limited to the western world. Japanese family crests date back to the eleventh century and have changed over time from their original usage among the aristocracy to emblems later used by the warrior class as they ascended to power at the end of the twelfth century. Emblazoned banners and flags were used to identify groups within the armies during war. During the Edo period (1603–1868), a time period that virtually overlaps the Age of Enlightenment going on in the west, family crests with symmetrical designs, primarily drawn from nature, became popular (Dower 1971; Matsuya 1972). The simplicity and directness of design and inspirational influence from nature comes from the strong influence of Shinto at this time. The use of these as ornamental emblems also became widely popular among the common populace rather than being restricted to upper society. With respect to living creatures, butterflies are the most common representatives. Matsuya (1972) illustrates over fifty butterfly emblems. Butterflies were a favorite among the aristocracy, almost certainly for their elegance (Dower 1971), and among young warriors as a symbol of the other side of their otherwise harsh life. Dragonflies are also found in some Japanese heraldry. These docile insects were known to the warriors by an old name meaning "insects of victory," a positive characteristic conveyed in the imagery of the designs on weapons and on family crests (Dower 1971). Even though Japan has native species of eusocial honey bees (*Apis cerana*) with similar characteristics to European honey bees, these insects are noticeably absent from Japanese heraldic crests.

DREAMS

Most people think of insects in a negative light and consider them to be associated with the negative aspects of life, including fear, disease, and dissimilarity from self (Kellert 1993). Such negative connotations can reside deep in the human psyche and thus insects can symbolically represent negative feelings and attributes in people's lives. Like the retention of irrational ideas ideally discarded by the understanding of the Enlightenment, deep-seated imagery of human experience of the past never really goes away and often manifests itself in our dreams (Klein 2012). In this way, insects have invaded our dreams and the interpretation of their presence and behavior in our dreams has received significant attention. Sometimes the imagery of dreams (either real or fictionalized) finds its way into works of literature and art and that in turn can shed light on the inner workings of the dreamer's mind or at least provide entertaining imagery. Lewis Carroll's Alice and her dream-based adventures in Wonderland and on the other side of a looking glass were written soon after the Enlightenment and introduced us to a few interesting bugfolk (e.g., a hookah-smoking caterpillar, rocking horse fly, snap dragon fly).

A particularly incredible case involves the dream, actually a vision experienced during a séance, by the English writer, poet, and painter William Blake (the same William Blake [Klonsky 1977] mentioned twice previously in this chapter). This incident was witnessed and recounted by his friend and collaborator John Varley, an English painter and astrologer (Klonsky 1977). The summoned spirit, described as the "Ghost of a Flea" and depicted in Blake's 1819 painting of the same name, is inspired by the morphology and behavior of a flea and symbolizes the dark, bloodthirsty excesses of some men as the flea told him (Blake) that "all fleas were inhabited by the souls of such men, as were by nature blood-thirsty to excess" (Butlin 1981). The hideous creature he saw, spoke to, and drew in his most famous vision-inspired painting, and accompanied by a close-up drawing of the head and mouth, is human in form, yet covered with a plated, scaly appearance similar to the external armor-like sclerites of a flea. Varley notes that Blake must surely be looking upon this sanguinary apparition in the room to have described him in words and illustration so vividly. Werness (2004) speculated that to know the features of a flea, Blake tapped into a deeper memory, that of Robert Hooke's engraving of a flea, published in his *Micrographia* of 1665 (Plate 34, and the cover image of this book). This rendering is an exquisite example of the microscopic discoveries to come in this new era of scientific discovery made possible by advances in microscopy (Ford 1993). Blake's spirit is thus the ghost of another flea from a more distant past.

The Enlightenment is touted as a time where the influence of scientific revolutions shed the light of recent discovery and reason on the world of ideas and society, and in turn led to the abandonment of many traditional ways of doing things. This scientific understanding was based on empirical observation and experimentation rather than on metaphysics and spirituality. People now should look at a world as behaving according to universal laws rather than being influenced by supernatural forces. And yet, human nature is such that novelty is foreign and uncomfortable, and tradition is familiar and hard to part with. Thus, the Age of Enlightenment, a time of reason and truth, was also a time of resentment of progressive change and a time where magic, myth, tradition, and fantasy continued to flourish. The study of insects played a large part in the advancement of science at this time. Insects also played a central role in the perpetuation of the old ways, both as symbolic tools of a disillusioned populace and simply as symbolic elements from the past.

CHAPTER SEVEN

Insects in Literature and Language

ANN M. MARTINEZ

Insects have long captivated readers' imaginations. Seeing them in day-to-day life, insects are both strange and familiar, terrifying and beautiful, but in literature these characteristics are further elevated to the point that "the butterfly fluttering in the narrow confines of [an] insectary can but suggest the beauty of the butterfly flitting unhampered against a background of rainbow-hued flowers in Keats' garden" (Eddy 1931a: 55). Literature heightens all of the insect's traits. Yet the interest they elicit from readers in texts has at times required justification. In *Natural History*, Pliny the Elder felt the need to validate his discussion of insects. He wrote, "nusquam alibi spectatiore naturae rerum artificio" (Nowhere else is nature's skill seen so well than in the smallness of insects) (qtd. in Brown 2006a: ix). In the span of years ranging from the Early Modern to the Romantic period, some of the most notable authors in English literature commonly used insects to fulfill various roles in their works. These roles are built on the fact that people develop a shared knowledge of insects, which authors then use to "make a point, elicit an emotion, or evoke a memory" (Berenbaum 2006: 3). For such critically observant authors like William Shakespeare, John Milton, and Percy Bysshe Shelley, the insects' malleability provides a range for actual and metaphorical associations in the literature.[1] When needed, these authors use insects to frighten on the one hand, and to model "proper" ethical, political, and cultural human behavior on the other.

"IF THE SUN BREED MAGGOTS": OF DEATH AND DARKNESS

Insects have long been associated with the darker side of nature. Since biblical times, they have evoked "frightening, dark, weird, and creepy sensations" because they remind humans of mortality through their presence in putrefaction (Berenbaum 2006: 6). One event that further solidified macabre connections between insects and death was the plague. Caused by the *Yersinia pestis* bacterium, the Black Death traveled from China to mainland Europe, entering through Venice, Italy, in 1347 and finally arriving in England in 1348. Some estimates place the death toll at one-third of the European population, exponentially slowing down the disposal of corpses because the living could not keep up with the burial of the dead. The accumulation of dead bodies made cadaver insects highly visible to the living, leading people to question their religious faith (Coutts 2006: 300–2). The medieval church maintained the teaching that after death a body could remain intact based on a person's sinless life. However, during the Black Death, people witnessed decomposition firsthand, and saw what appeared to be the metamorphosis of the human body into insects—a transformation that was deemed evil (Coutts 2006: 305). Eventually, people's views changed to a degree, and by the seventeenth century, much of the educated population was certain "that insects were produced exclusively from corruption, spontaneously generated from rotting leaves, dead bodies, all sorts of putrefaction" (Manley 1961: 400). The issue at hand then became to explain the presence of these insects: either God created them, they created themselves from death, or they were a product of original sin. Some authors successfully incorporated speculative answers about the origin of insects into their literary works, reflecting society's views.

The seemingly spontaneous generation of insects is central to one of Shakespeare's most well-known scenes of madness in *Hamlet*.[2] Back in the Early Modern period, a common belief in England was that the sun helped generate insects in a decomposing body (Patterson 1838: 29).[3] This view, however, was not necessarily a negative one. Shakespeare alludes to this belief through Hamlet's wordplay in the scene with Polonius. Hamlet states, "For if the sun breed maggots in a dead dog, being a | good kissing carrion—have you a daughter?" (2.2.179–80). After Polonius responds that he does have a daughter (Ophelia), Hamlet adds, "Let her not walk i'th' sun. Conception is a blessing, | but as your daughter may conceive, friend—look to't" (2.2.182–3). The exchange not only points toward the sun's life-generating powers by "breeding" maggots from a carcass, but to the central state in the generation of life. According to the early modern physician Paracelsus, "putrefaction represents the end of corruption and the beginning of generation; it thus marks the transition from death to birth" (Hankins 1949: 510). This brings us back to the line "being a good kissing carrion."[4] John Hankins proposes that:

> [i]f the sun is the kisser, then the phrase means "good for receiving the sun's kiss"; i.e., the corrupted flesh is already predisposed to the breeding of life from the impulse of the sun's rays and for that purpose is superior to living flesh.
>
> (Hankins 1949: 514)

Rather than seen as the foul destruction of the human body, as it once was, the appearance of maggots becomes an instrument for Shakespeare to emphasize new life over death, and he quickly underscores his point by a reference to Ophelia and pregnancy. Hamlet warns Polonius to keep his daughter indoors, away from the sun, due to the myth-based belief that the sun's rays can cause a woman to conceive.[5] Hamlet's desire to keep Ophelia from conceiving is not to protect her chastity but because he wishes to limit human births based on his own regret at having been born, as he later says to her:

> Get thee to a nunnery! Why wouldst thou be a breeder of sinners? I am myself indifferent honest, but yet I could accuse me of such things that it were better my mother had not borne me.
>
> (3.1.119–22)

The two scenes become a reflection of Hamlet's state of mind toward his own actions; however, the presence of the maggot-infested body is emphasizing life rather than death in this instance.

Insects in literature easily transitioned from being associated with mortality to becoming ominous creatures of evil. *Macbeth*, one of Shakespeare's darkest plays, includes a number of passages where insects are used to underscore evil portents. Immediately after Macbeth murders Duncan, Lady Macbeth hears "the owl scream and the crickets cry" (2.2.15), even though in literature the cricket's chirp has long been associate with "gaiety and mirth" (Patterson 1838: 100). Shakespeare's use of "cry" instead of "chirp" emphasizes the portentousness of the moment announcing the betrayal and gruesome murder of a caring and good king, and appears in stark contrast with his use of crickets in other plays. For example, in *Pericles*, after the wedding feast, Gower, appearing in the role of the chorus, describes the surrounding scene of napping guests adding, "And crickets sing at the oven's mouth | Are the blither for their drouth" (3.0.7–8). Here, the crickets' singing is indicative of their happiness at being dry and warm, echoing the atmosphere of the scene. Similarly, in *The Winter's Tale*, crickets are mentioned during the only happy encounter on stage between Hermione and her son, Mamillius. As the boy prepares to entertain his mother with a ghost story, he says "I will tell it softly; | Yond crickets shall not hear it" (2.1.31–2). Even though he is still a child, Mamillius exhibits his keen observation in noting that crickets fall silent when humans speak near them (Engeln 1916: 584).[6]

Crickets, however, are not the only insects Shakespeare occasionally links to humanity's dark moments. As Macbeth laments to Lady Macbeth that Banquo and Fleance still live, he says,

> … … … … … … … … …. they are assailable.
> Then be thou jocund. Ere the bat hath flown
> His cloistered flight, ere to black Hecate's summons
> The shard-born beetle with his drowsy hums
> Hath rung night's yawning peal, there shall be done
> A deed of dreadful note.
>
> (3.2.38–43)

Macbeth speaks of the beetle as if it were in service to Hecate, the Greek goddess often associated with witchcraft and necromancy. While this insect is meant to announce the beginning of night, its humming will instead serve to announce that "a deed of dreadful note" has been accomplished, which in this case refers to the murder of Macbeth's loyal friend and his young son. The beetle's color adds consistency to the scene since everything mentioned is black or dark: the bat, Hecate, night-time, and the state of sleep—the latter often thought of as "death's second self" due to sleep's dark state of slumber. While the imagery itself is clear in this passage, Shakespeare's "shard-born beetle" is likely the bard's most academically debated insect. The question lies with the definition of "shard" (whether it refers to the creature's body, its wings, or its origins from dung), a point discussed by scholars since 1743.[7] In all of Shakespeare's works, there are only three mentions of the shard-born beetle, appearing in plays ranging from 1606 to 1609. Aside from the lines previously quoted from *Macbeth*, this particular insect is found in *Antony and Cleopatra*, where Enobarbus, speaking of Lepidus' kowtowing to Antony and Octavius says, "They are his shards and he their beetle" (3.2.19), while in *Cymbeline,* it appears when Belarius analogizes the contrast of life in their cave with life at court, saying, "And often to our comfort shall we find | The sharded beetle in a safer hold | Than is the full-winged eagle" (3.3.19–21). The use of the beetle in these two instances is quite different from its appearance in *Macbeth*; the ominousness is gone in *Antony and Cleopatra*, its place taken by an insult, while in *Cymbeline* the lowly beetle is safer in its abode than the high soaring eagle in its nest.

In 1778, George Tollet first proposed that the sharded beetle refers particularly to the beetle "born in cow-dung," based on the fact that in the Staffordshire region the term *cow-shard* commonly refers to cow-dung (qtd. in Patterson 1838: 67). The association of the beetle with excrement grew due to the popularity of Aesop's fables during the seventeenth century, particularly "De Scarabeo et Aquila," in which the beetle throws excrement at Zeus so that the god will stand up and drop the eagle's eggs that he guards; similarly, Aristotle, Pliny, Plutarch, and Isidore wrote that the beetle was born of excrement, which in turn was supported through "the natural history of the beetle's dependence on excrement" (Billings 2005: 440, 445). Although many editors still note the different glosses for "shard-born" (as both relating to wings or dung), Timothy Billings argues convincingly for cow-dung as the original intent. He notes that it was not until the nineteenth century when the *Oxford English Dictionary* listed a definition of "shard" as "the hard wing case of a coleopterous insect." This appears to be an instance of scholars erroneously defining a term based on what they thought was Shakespeare's use and due to his reputation and authority over the English language (2005: 446).

"LIKE A WORM I' THE BUD": OF DAMAGE AND TORTURE

While authors from the sixteenth and seventeenth centuries relied on the insect for darker symbolism, the more common and realistic link was simply to physical damage. In many literary instances, insects, in order to survive, unintentionally destroy that which humans value. For example, in *Hamlet*, when Laertes warns his sister Ophelia to be wary of Hamlet's advances, he analogizes the situation with that of the worm that must find nourishment in new sprouts. Laertes says, "Virtue itself scapes not calumnious strokes, | The canker galls the infants of the spring | Too oft before their buttons be disclosed, | And in the morn and liquid dew of youth | Contagious blastments are most imminent" (1.3.37–41). In this case, the canker, or worm, does what it needs to do for its own sake, but in doing so damages new growth. Shakespeare returns to the analogy of destruction in *Twelfth Night*, when Viola speaks to Duke Orsino about the ills of unrequired love, saying, "She never told her love | But let concealment, like a worm i'th' bud, | Feed on her damask cheek" (2.4.107–9).[8] In other literary instances, however, insects are themselves the instruments of torture. Shelley, in his satirical play *Oedipus Tyrannus or Swellfoot the Tyrant*, uses the gadfly, or horsefly, as a weapon. As Purganax and Mammon, both advisors to Swellfoot, discuss ways of torturing Queen Iona, Purganax speaks of the gadfly as one of three creatures he pulled out of a hellish cavern "full of ugly shapes" (1820: I.150). He traces the origin of the gadfly to having been summoned by

God “out of the mountains | Of utmost Aethiopia, to torment | Mesopotamian Babylon” (I.154–5). Purganax describes in detail the creature’s horridness:

The beast
  Has a loud trumpet like the scarabee,
  His crookèd tail is barbed with many stings,
  Each able to make a thousand wounds, and each
  Immedicable; from his convex eyes
  He sees fair things in many hideous shapes,
  And trumpets all his falsehood to the world.
  Like other beetles he is fed on dung—[9]
  He has eleven feet with which he crawls,
  Trailing a blistering slime.

(I.156–65)

It is not only that the gadfly can sting, injecting countless injuries that are incurable, but through its large compound eyes the world’s beauty is distorted. Because the gadfly can only see an unsightly world, its surroundings always appear threatening. As such, the gadfly mindlessly attacks the ugliness around it, leaving no opportunity for humans to try to reason with such a creature.

“AS FLIES TO WANTON BOYS”: OF INDIVIDUALITY AND HUMANITY

Recently, social ecologist Stephen R. Kellert has identified five factors that affect people’s dislike of insects. Two of these factors are of particular note in the study of literature: a) the large number of insects challenges the human focus on identity and individuality; and b) insects are thought to be mindless and have no emotions, like Shelley’s gadfly (1993: 57–8).[10] More visible insects, those who occupy shared spaces with humans, can easily be associated with their multiplicity, like ants, for example, whose numbers often bring to mind an army. We as humans have come to value identity and uniqueness in ourselves; thus, the large array of seemingly identical and easily replaceable insects can trigger feelings of discomfort and insecurity by analogy. In Shakespeare’s *King Lear*, Gloucester speaks of insects to lament the degree to which humans are expendable when he says, “As flies to wanton boys are we to th’ gods: | They kill us for their sport” (4.1.37–8). Although having a sense of identity has been a constant through the periods of human history, it is during the Early Modern period when a concern for individuality came more prominently to the forefront of cultural conversations. Deemed “self-fashioning” by modern scholars, the concept refers to the proliferation during the sixteenth and early seventeenth centuries of cultural norms and literature that harkened toward "a distinctive personality, a characteristic address to the world, a consistent mode of perceiving and behaving" (Greenblatt [1980] 2005: 2), and one that, as Edmund Spenser called for in *The Faerie Queene*, was modeled on "vertuous and gentle discipline" ([1590] 2014: 402). However, in aspiring for more perfectly behaved people, sometimes the insects' lack of individuality can be favorable. In Shelley's long poem *Queen Mab*, where he proposes a future utopian society in which people work with nature and not against it, he removes the sense of individuality and instead praises the analogous insect when he describes the view of people on Earth from the heavens as "The thronging thousands to a passing view, | Seemed like an anthill's citizens. | How wonderful!" ([1813] 1831: 18). For someone whose poetry always glances attentively at nature, Shelley calls on the image of the structure and organization of the ant when he uses the term "citizen," doubly humanizing the ants while slightly dehumanizing people.

In fact, humanizing insects is a common trope in literature as it allows authors to create nuanced connections. To do so, authors have to break down some of the readers' assumptions because insects are "feared for their fierce martiality, their consuming appetite, and their mobbing numbers" (Brown 2006b: 38). They possess superior physicality, considering their dimensions, they compete with humans for food, and they outnumber humans. More significantly, they "exist beyond our capacity for language, specifically unnameable by anyone but themselves" (Brown 2006a: xii). Humans cannot reason with insects because we do not speak the same language, and never will. However, even with such a divide it is still possible to create different commonalities so that humans can understand insects and relate to them in various ways. During the late sixteenth and early seventeenth centuries in England, both apiculture and sericulture became very popular amongst the gentry, leading to the publication of a high number of manuals. These texts invariably anthropomorphize bees and silkworms in relation to each other and in relation to human society. Bees were seen as morally superior to the silkworm due to the bees' "work-ethic": bees produce honey, which is healthy and nutritious, and wax, used to produce candles for illumination, while the silkworm only produces the material for expensive bodily adornment; furthermore, bees are hardworking, starting their chores early in the day, while silkworms simply lounge around feeding (Olbricht 2006: 223).[11] Early modern manuals advised beekeepers to lead sexually pure lives and to follow a personal hygiene regimen so as to be accepted by the bees and avoid their stings (Olbricht 2006: 236). In his 1623 edition of *The Feminine Monarchie*, Charles Butler instructs beekeepers to gain the bees' love by being "chaste, cleanly, sweet, sober, quiet, and familiar" (1623: C2r). The instructions Butler gave regarding silkworms was somewhat different, as he emphasized cleanliness and sweet smells over everything else. As Olbricht explains, Butler equated the insects with the upper classes, indicating that silkworms possessed "a keen sense of smell, equivalent to that of the nobility, and [were] offended by dirt and the poorer classes" (2006: 236). Butler did not limit his discussion to the behaviors exhibited by those around bees or silkworms; he also modeled a perfect society on the insects' organization. He asserted unequivocally that "Bees abhorre as well Polyarchie, as Anarchie, God hauing shewed in them vnto men, an expresse patterne of A PERFECT MONARCHIE, THE MOST NATVRAL AND ABSOLVTE FORME OF GOVERNMENT" (1623: B2r). In fact, he is not alone. The bee, as a political model, has long been popular in literature.

"FOR SO WORK THE HONEYBEES": OF MORALITY AND MONARCHY

Metaphors of insects are based on a montage of the human individual and society. Christopher Hollingsworth sees the insect metaphor as a product of people's acceptance of the Great Chain of Being; he explains, "[b]ecause we aspire to an image of its politics, we saw the natural beehive as kin. Rhetoric not biology, analogy rather than blood, tied us to this living allegory, which signified visually in a language that all could understand" (2001: 152). The socio-political structure of bees becomes a key point of argumentation in Shakespeare's *Henry V*.[12] As Karen L. Edwards mentions, "[p]airing ants and bees as complementary models of industry and government is a commonplace of classical and Renaissance literature" (1999: 129). As the king and his advisors discuss Henry's claim to the French throne and the impending war, the Earl of Westmorland brings up the various social classes in England and their differing functions. To this, the Archbishop of Canterbury responds in agreement with the following:

> Therefore doth heaven divide
> The state of man in diverse functions,
> Setting endeavor in continual motion,
> To which is fixèd, as an aim or butt,
> Obedience. For so work the honeybees,
> Creatures that by a rule in nature teach
> The act of order to a peopled kingdom.
> They have a king and officers of sorts,[13]
> Where some like magistrates correct at home,
> Others like merchants venture trade abroad;
> Others like soldiers, armèd in their stings,
> Make boot upon the summer's velvet buds,
> Which pillage they with merry march bring home
> To the tent-royal of their emperor,

Who busied in his majesty surveys
The singing masons building roofs of gold;
The civil citizens kneading up the honey;
The poor mechanic porters crowding in
Their heavy burdens at his narrow gate;
The sad-eyed justice with his surly hum
Delivering o'er to executors pale
The lazy yawning drone.

(1.2.183–204)

Canterbury presents a harmonious society centered around a monarchical figure. Several of the social tiers are seemingly content in their roles—the soldiers are "merry," the masons sing as they work, and the citizens are "civil." The only ones unhappy are those who must enforce punishment, the justices, over the "lazy" drones. If people do their duty toward their king, stay within their station and work diligently, the nation will be all the better for it.[14]

For John Milton, an outspoken disestablishmentarian and strong proponent of regicide against Charles I, bees do not afford an ideal model of government *precisely* because it is monarchical. Unlike authors who came before him, like Virgil and Dante, who placed the hive in Heaven, Milton placed it in Hell (Hollingsworth 2001: 85–6). After Satan and the other Fallen Angels wake up in Hell, they build a new palace, Pandemonium. As they proceed to congregate inside the palace for the first time, Milton describes their movement:

The gates
And porches wide, but chief the spacious hall
… … … … … … … … … … … … … … .....
Thick swarmed both on the ground and in the air,
Brushed with the hiss of rustling wings. As bees
In spring time when the sun with Taurus rides
Pour forth their populous youth about the hive
In clusters, they among fresh dews and flow'rs
Fly to and fro or on the smoothèd plank,
The suburb of their straw-built citadel,
New rubbed with balm, expatiate and confer
Their state affairs, so thick the airy crowd
Swarmed and were straitened, till the signal giv'n.

([1667] 2005: I.761–76)

The Fallen Angels, now transformed into Demons, are directly analogized to bees in their behavior, their movements, their sound. They swarm toward their hive just as drone bees might upon returning at the end of day, summoned by their monarch—in this case, Satan. In *Henry V*, the hive is an example of order and control. In *Paradise Lost*, it is literally hellish. The hive becomes a metaphor that represents the contemporary views of society as reflected by each of the authors (Hollingsworth 2001: 99).

Milton both critiqued bees and focused on them more than any other insect. In all of his poetry, he mentions a total of seven insects fourteen different times.[15] Of these, the bee gets the most mentions at five, and is described as one of the first insects created by God. In fact, in attributing insects to God, Milton takes a stand on a much-contested issue. As mentioned previously, the origin of insects was uncertain because it carried with it the baggage of good, evil, death, and original sin. Milton clarified the issue in *Paradise Lost* by explaining that insects were created on the last day. The description is a positive one, with the typically dark and often slimy associations removed. He writes,

> At once came forth whatever creeps the ground,
> Insect or worm. Those waved their limber fans
> For wings and smallest lineaments exact
> In all the liveries decked of summer's pride
> With spots of gold and purple, azure and green.
>
> ([1667] 2005: VII.475–9)

Milton echoes Pliny the Elder in his praise of the craft and detail found in small insects. Even the colors mentioned are those highly prized for their brightness. From the physical appearance of the insects, Milton further directs the reader's gaze to narrow in on the behavioral qualities of the ant and the bee:

> First crept
> The parsimonious emmet, provident
> Of future, in small room large heart enclosed,
> Pattern of just equality perhaps
> Hereafter, joined in her popular tribes
> Of commonalty. Swarming next appeared
> The female bee that feeds her husband drone
> Deliciously and builds her waxen cells
> With honey stored.
>
> (VII.486–92)

Just as the beekeeping manuals popular earlier in his century do, Milton also praises the work ethic of these insects. He, too, anthropomorphizes them speaking of the ant's frugality, good-heartedness, and magnanimity. In comparison, the bee falls short, as he sees it, because her main attribute is simply to attend to her husband and store food for him, a point that Milton extends in other writings to the relationship of the people to their king.

"WHEREFORE, BEES OF ENGLAND, FORGE | MANY A WEAPON": OF GOVERNMENT AND REBELLION

In *Eikonoklastes*, Milton justifies the execution of Charles I, which had taken place nine months prior, by using insects in his analogy. In doing so, he responds directly to the *Eikon Basilike*, the supposed but never confirmed spiritual autobiography of Charles I, in which the narrator states that "Bees will gather honey where the Spider sucks Poyson," meaning that the king's true followers will always select material from the autobiography that is true to the king's memory, while his detractors will zero in on the problematic passages (Jacobs 2015: 183). Milton responds to this with bees. He writes that,

> They who can pick nothing out of [the autobiography] but phrases shall be counted *Bees*: they that discern fur[th]er both there and here that *constancy to his Wife* is set in place before Laws and Religion, are in [his followers' judgement] ... no better then [*sic*] *Spiders*.
>
> (3.541)

According to Hollingsworth, Milton did not want to conflate people with bees because people have free will, something he valued greatly (2001: 96). The words he directs toward the *Eikon Basilike* are a harsh critique on his part precisely because they hint at the lack of will found in the bee, granting the spider reason and agency.

Shelley, considered by some to be a poet with a "marked fondness for winged insects" (Hughes 1918: 290), does not condemn bees for their monarchical tendencies but instead praises their productivity.[16] In his "Song: To the Men of England," Shelley draws a parallel between bees and the upper and working classes in England. Calling on the "men of England," here specifically the working class, Shelley asks them to consider what they do for the landed English lords whom he has equated to drones:

> Men of England, wherefore plough
> For the lords who lay ye low?
> Wherefore weave with toil and care
> The rich robes your tyrants wear?
> Wherefore feed and clothe and save
> From the cradle to the grave

Those ungrateful drones who would
Drain your sweat—nay, drink your blood?

(Shelley [1819] 2010: 77–8)

Shelley, never one to shy away from expressing his views on religion or on social class injustice, equates the lords with drones, feeding off the work of others and seeming more like blood-sucking vampires than male bees.[17] The English working class, for Shelley, keeps the country running just as the worker bee does for the hive. He adds,

Wherefore, Bees of England, forge
Many a weapon, chain, and scourge,
That these stingless drones may spoil
The forced produce of your toil?
... ...
Sow seed—but let no tyrant reap:
Find wealth—let no imposter heap:
Weave robes—let not the idle wear:
Forge arms—in your defence [*sic*] to bear.

(78)

His call to civil disobedience is devoid of any associations with the monarchy, and thus Shelley is not inciting rebellion against the English Crown. Instead, he is asking the working class to no longer support the gentry as they do, and to instead work for themselves. Poetically speaking, it is a stirring comparison, so much so, in fact, that the Chartists of the 1840s took up Shelley's poem as one of their movement's songs (Behrendt, qtd. in Shelley 2010: 77).

Entomologically speaking, it is not a viable option. If the worker bees cease to feed and care for the drones, the drones will not join the workforce but simply die, as will the hive as a whole with no males available for reproduction. Shelley, in truth, was not overly concerned with scientific accuracy. As Lloyd Jeffrey explains,

> to a man like Shelley the formal identities of living things mean little as against their imaginative embodiments. The mind that compares its poetic career to the flight of a "silken-wingéd fly" ... is not likely to wonder whether his conceit was inspired by suborder *Zygoptera* or by suborder *Anisoptera*.
>
> (1976: 104)

The poet's use of insects in his work has been labeled "impressionistic entomology" due to the fact that it is based on a combination of sources: primarily personal observation, English folklore, classical literature, and the Bible, and, secondarily, a likely handful of scientific texts (Jeffrey 1976: 115). One might also call it poetic license.

In the period spanning 1600 to 1820, Shelley was very likely not the only "impressionist entomologist." Authors like Shakespeare and Milton also gave insects various roles to play in the literature of their period. At times the harbingers of death, at others the models for proper human behavior, insects were ever present in texts frightening, inspiring, or prompting audiences and readers alike to think of their own humanity. Always adaptable, "[i]nsects are chameleons changing their color to blend with the tones of the poet's mood" (Eddy 1931b: 156), and so reflect a generation's sensibilities toward and knowledge of these same creatures.

CHAPTER EIGHT

Insects in Art

JENNIFER ANGUS, KRISTEN E. REITER, AND MATTHEW S. LEHNERT

The Age of Enlightenment signaled a shift in the way natural phenomena were understood. Philosophers such as Francis Bacon and René Descartes, and the natural philosophers, now recognized as scientists, most notably Isaac Newton, argued that truth was to be found through observation, experimentation, and analysis, a process we now know as the scientific method. While the church remained the moral authority, no longer could cataclysmic events such as a volcanic eruption or lightning strike be attributed to a sign of God's will. These wonders could be explained through science—astronomy, chemistry, and the emerging fields of fulminology, geology, and physics. New technology, in the form of the microscope, heralded enormous advances in biology, as microorganisms were revealed with magnification abilities that far exceeded those of the human eye. Renowned artists of the day were called upon to illustrate this uncovered world. With a new focus on organisms at the microscale, it is not surprising that the secret lives of insects would be explored by pioneers, such as Maria Sibylla Merian—a painter, illustrator, and a woman far ahead of her time, who would make valuable contributions to the field of entomology. Scientific illustration entered its golden age.

The dominant art movement during the Age of Enlightenment was the Baroque style, characterized by grandeur, exuberance, and ornate detail. The later Baroque period, known as Rococo, was so over-the-top and highly theatrical that it would ultimately be deemed both excessive and trivial. Supported by the Catholic Church, the Baroque movement was seen as a means to counter the austerity of Protestant art and architecture. While the authority of science and reasoning was to usurp the church's wide-ranging influence, its effect upon the art world was less dramatic. The arts were dependent upon patrons who were guided by the expectations and traditions of social standing as well as their own personal taste. However, the advent of the printing press in the fifteenth century allowed scientists to disseminate their research in the form of books and pamphlets. As such, there was a demand for realistic scientific illustrations to accompany their findings.

Early depictions of exotic creatures, large mammals in particular, are bizarrely fanciful, in part, because artists had never seen actual specimens. They relied on verbal and written descriptions, as well as a healthy imagination to illustrate the animals they were commissioned to draw. Insect illustrations, however, have been relatively accurate throughout the ages since specimens are relatively small and thus quite portable. The few exceptions that exist, such as the "Locust of Kalisz" (Figure 8.1) by General Joachim Daniel, dating from 1749, were likely intended as caricatures or truly meant to represent a beast of legend (Shuker 2017).

FIGURE 8.1 The "Locust of Kalisz" was found within a scrapbook that was prepared by friends of Polish General Joachim Daniel von Jauch on the occasion of his birthday and likely intended as a humorous depiction of the general himself.

The dragon-type creature found in the *Volatilium Varii Generis Effigies* series (Figure 8.2) possesses insect qualities—six legs, four wings, moth-like antenna, and a spiral-type proboscis found in lepidopteran species. Whether it is a mythical beast or an inexpert rendering of an unknown winged creature, it is not nearly as frightening as a true menace, the flea (Figure 8.3), so accurately illustrated by Robert Hooke (1635–1703) in his book *Micrographia: or Some Physiological Descriptions of Minute Bodies Made by Magnifying Glasses. With Observations and Inquiries Thereupon* published in 1665. Hooke's flea is enlarged from its actual size to reveal a monster of nightmarish proportions. Yet observed through the lens of a microscope, Hooke admiringly wrote:

> The strength and beauty of this small creature, had it no other relation at all to man, would deserve a description ... the Microscope manifests it to be all over adorn'd with a curiously polish'd suit of sable Armour, neatly jointed, and beset with multitudes of sharp pinns, shap'd almost like Porcupine's Quills.
>
> (Hooke 1665)

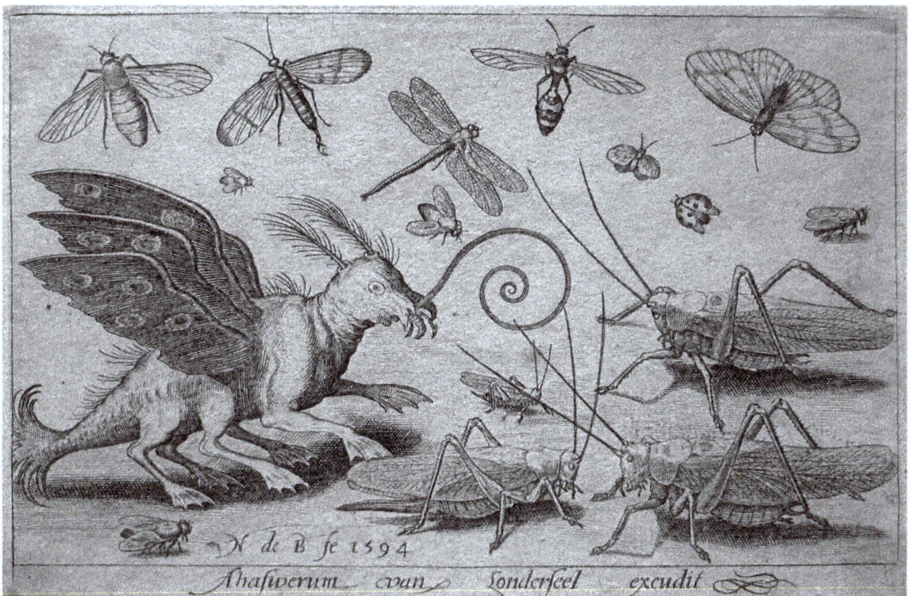

FIGURE 8.2 Nicolaes de Bruyn (1571–1656), Flemish engraver, 1594, from *Volatilium Varii Generis Effigies* (Pictures of Flying Creatures of Varied Kinds).

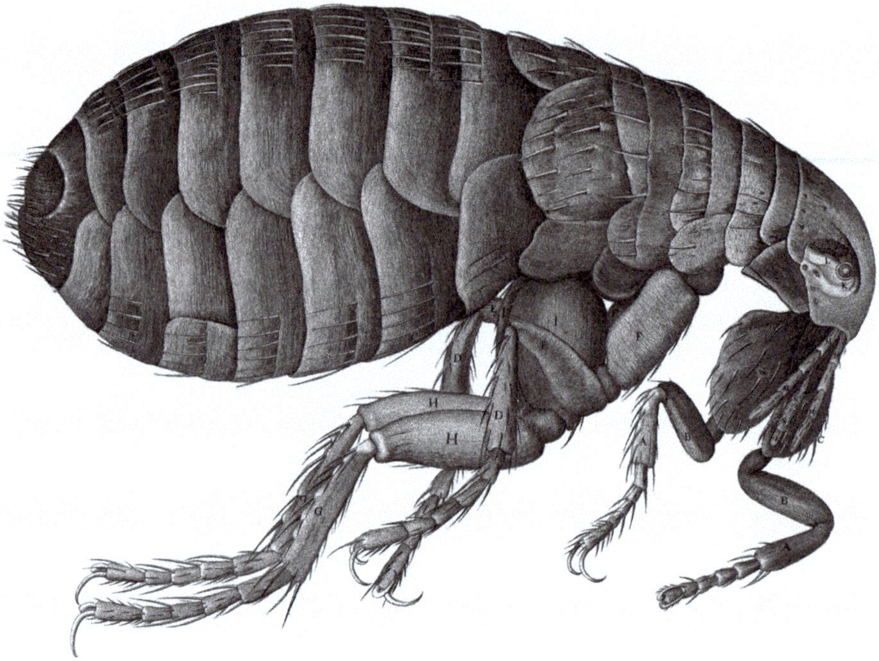

FIGURE 8.3 *Micrographia*, Schem. XXXIV, "An illustration of a flea." Image acquired from the National Library of Wales. Wikimedia Commons.

The purpose of this chapter is to provide examples of insects depicted in artwork during the Age of Enlightenment, which will include examples from scientists, amateur entomologists, and other artists from Europe.

ENTOMOLOGISTS AS ARTISTS

A remarkable man with wide-ranging interests, Robert Hooke (1635–1703), dubbed the "Leonardo of London," was a man of multiple talents and one of the few scientists capable of illustrating his subjects himself. As a teenager, he apprenticed with Peter Lely (1618–1680), a leading portrait painter in the court of Charles II. It has been suggested that had Hooke not been such a cantankerous individual, most famously feuding with Isaac Newton, he might well have been considered the leading experimental scientist of his day. He was a polymath with skills in architecture, artistry, astronomy, biology, cartography, chemistry, geology, and mechanics. For forty years, Hooke held the position of Curator of Experiments at the Royal Society. Established in 1660, the Royal Society is the oldest national scientific institution in the world. Hooke collaborated and corresponded with contemporaries, such as Christopher Wren (1632–1723), best known as the architect of St. Paul's Cathedral in London, Robert Boyle (1627–1691), considered the first modern-day chemist, and Isaac Newton (1643–1727), ultimately Hooke's nemesis for he felt that due credit was not received for research that assisted Newton and catapulted him to fame.

Micrographia was the first book published by the Royal Society and it enjoyed tremendous success beyond the scientific community. Simply put, it was revolutionary. Samuel Pepys, a member of parliament at the time wrote, "Before I went to bed, I sat up till 2 a-clock in my chamber, reading of Mr. Hooke's Microscopical Observations, the most ingenious book that I ever read in my life" (Cobb 2006). Chapters devoted to observations upon a sting of a bee, the feet of a fly, or the eggs of a silkworm revealed a hidden world that was both breathtaking and fascinating. In his preface he declared,

> it is my hope, as well as belief, that these my Labours will be no more comparable to the Productions of many other Natural Philosophers, who are now every where busie about greater things; then my little Objects are to be compar'd to the greater and more beautiful Works of Nature, A Flea, a Mite, a Gnat, to an Horse, an Elephant, or a Lyon.
>
> (Hooke 1665)

Hooke's "little Objects" were mundane, yet his illustrations were nothing but extraordinary. His fly's eye (Figure 8.4) is so detailed, so threatening looking, and ultimately so alien that it is understandable why there were skeptics. The illustration of a single specimen might occupy a full page, while to other specimens he devoted a double page spread, thus creating a truly monstrous sight of epic proportions. One of the most interesting aspects of Hooke's illustrations that may not be obvious to a modern-day reader is how groundbreaking was the use of a circular frame to demark the sight line through the microscope. This view would have been an entirely new experience for the reading public. Hooke wrote,

> By the means of Telescopes, there is nothing so far distant but may be represented to our view; and by the help of Microscopes, there is nothing so small as to escape our inquiry; hence there is a new visible World discovered to the understanding.
>
> (Hooke 1665)

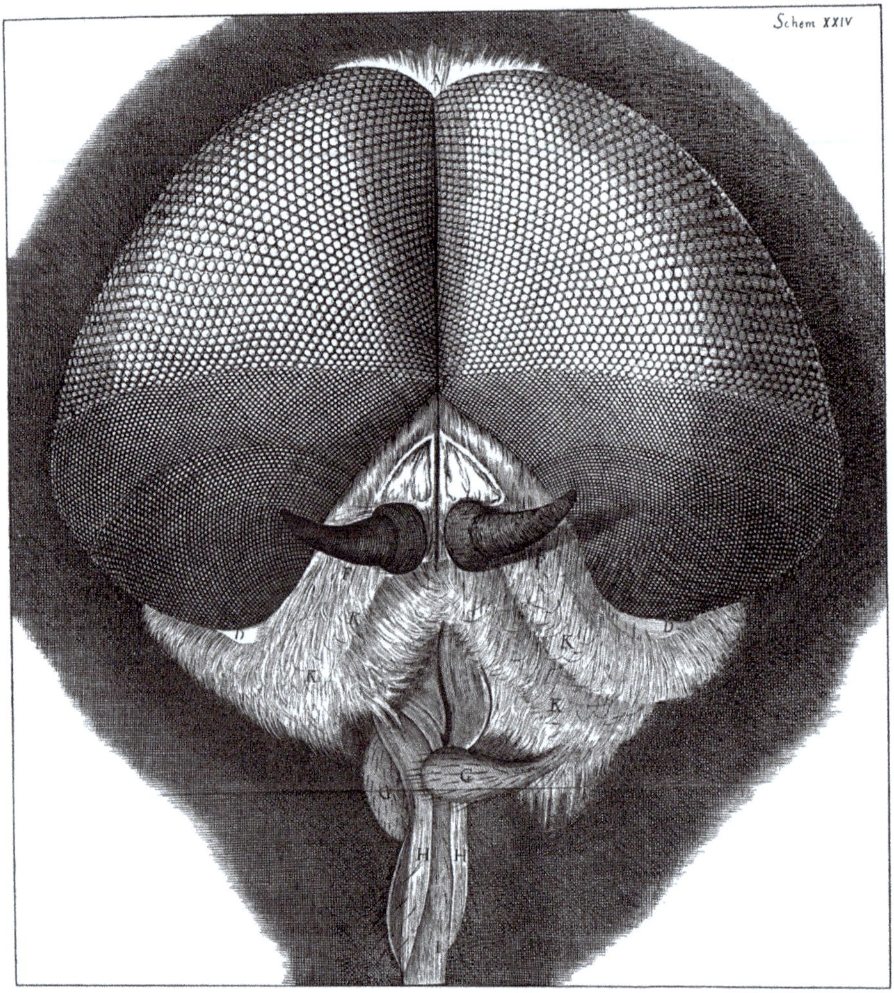

FIGURE 8.4 *Micrographia*, Schem. XXIV, "Of the Eyes and Head of a Grey drone-Fly." Wikimedia Commons.

Although Robert Hooke performed outstanding artwork based on what he observed through a microscope, the artistic contributions by Dutch scientist Jan Swammerdam (1637–1680) forever changed the field of entomology. Swammerdam had an early interest in natural history and studied medicine at the University of Leiden, but his true passion was with insects. His influences on entomology are far too extensive for the scope of this chapter, but some of his achievements included determining that the different stages of an insect's metamorphosis are variations of the same animal, that insects have an internal anatomy, and demonstrating that the "king bee" was in fact a queen

(Enenkel and Smith 2007). Despite a lack of support from his father (who wanted Swammerdam to focus his time on medicine), Swammerdam published his famous studies of insects in the book *Historia insectorum generalis ofte Algemeene verhandeling van de bloedeloose dierkens* (The General History of Insects, or General Treatise on Little Bloodless Animals) in 1669. His illustrations are meticulous. Swammerdam used microscopy to reveal the secret biology of insects, but it was his ability to accurately illustrate his findings that served as a vehicle to support his science. For instance, it was in his book *Biblia naturae* that Swammerdam provided evidence that the "king bee" is a female, because of the eggs found among the internal anatomy of her reproductive organs (Munz 2016).

Entomological illustrations, such as those pioneered by Swammerdam, increased in popularity during the remainder of the Age of Enlightenment. Moses Harris, a career entomologist, collected and illustrated insects, especially lepidopterans. His *The Aurelian or Natural History of English Insects; namely, Moths and Butterflies, together with the Plants on which they Feed* (1766) features illustrations of the appearance and markings of adult butterflies and moths as well as depictions of the collection process, such as how to catch and prepare insect specimens, which were illustrated in detail (Weiss 1926). Harris also was the artist behind most of the colored engravings featured in Dru Drury's three-volume *Illustrations of natural history, wherein are exhibited figures of exotic insects* (1770–1782), which depict and describe over a thousand foreign insect species (Cockerell 1922).

Jacob Hubner was also a mass-producer of insect illustrations, publishing well over one thousand engravings throughout his career. His propensity for drawing, obvious to his peers as a young boy, inspired Hubner to become an etcher. In his early twenties he developed an interest in natural history. The combination of these two passions led Hubner to create numerous entomological illustrations. By the age of twenty-three, Hubner attempted his first engraving depicting insects—a copper plate with moths and butterflies. Two years later, he began publication of an illustrated work describing the early stages of lepidopteran development, *Beiträge zur Geschichte der Schmetterlinge* (1786). Hubner's illustrated plates in this work are generally regarded as very well executed. For Hubner's monumental *Geschichte europäischer Schmetterlinge*, another publication dedicated to illustrating lepidopteran life stages, he engraved over five hundred plates, some of which continued to be published following Hubner's death. A companion publication featuring illustrations of adult European Lepidoptera, *Sammlung europäischer Schmetterlinge* (1796), featured seven hundred plates by Hubner at the time of his death, with additional illustrations published posthumously. Huber's focus at the turn of the nineteenth century shifted to the description and depiction of exotic lepidopterans in his *Sammlung exotischer Schmetterlinge* (1806) in which he published almost five hundred additional engraved plates over a span of twenty-two years. Hubner engraved over a hundred plates for his 1808 supplement to *Sammlung exotischer Schmetterlinge*, *Erste Zuträge zur Sammlung exotischer Schmetterlinge*, illustrating almost four hundred exotic species of Lepidoptera. Following the death of his wife in 1820, Hubner continued to publish a small number of entomological engravings each year but did not produce any additional major works and passed in 1826 (Hemming 1937).

Like Hubner, contemporary Edward Donovan produced hundreds of illustrations of insects. Well known for balancing empiricism with aesthetics, his accurate depictions have been described as bridging the gap between the focus on art in eighteenth-century illustrations and on science in those of the nineteenth century. His career in scientific illustration began with a focus on plants—the highly accurate true-to-size and magnified illustrations of Donovan's *The botanical review* (1789–1790) became the major appeal of the work. His volumes on insects, *The Natural History of British Insects 1792–1813*, feature over 560 illustrations of insects drawn from live specimens to create the most credible depictions of each species following the precedent set by Maria Sibylla Merian (see later in this chapter) (Gibson 2017).

ARTISTS INSPIRED BY INSECTS

Although there were many entomological illustrators during the Age of Enlightenment, Joris and Jacob Hoefnagel, Flemish artists in the late sixteenth and early seventeenth centuries, were traditional artists who regularly depicted insects in their works. Joris Hoefnagel (1542–1600) was a merchant and a famed painter with particular skill when painting miniatures. Jacob Hoefnagel (1573–1632), Joris's eldest son, trained as a painter and was heavily involved in politics (Vignau-Wilberg 2017). Joris began painting as a child, expressing his interest in the world and the organisms that inhabit it through his art. Joris had extensive knowledge of natural history and intended his works to not only be aesthetically pleasing, but to also provide scientific information (Figure 8.5). His paintings addressed topics that ranged from religion and mythology, to topography, to natural history, including insects. Joris was not only a painter of insects, but often corresponded with others, such as Thomas Muffet, who shared a fascination with the biology of these creatures. In his four-volume work *The Four Elements* (1591–1599), Joris assigned each animal he portrayed to one of the four elements. He was particularly intrigued by those he assigned to the element of fire—insects. *Ignis: Anamalia Rationalia et Insecta* depicted mayflies, mantids, beetles, lepidopterans (including pupae), craneflies, earwigs, and bees, among other insects.

At the time of *Ignis*'s publication, the major early works on entomology, such as Muffet's *Insectorum sive Minimorum Animalium Theatrum*, had not yet

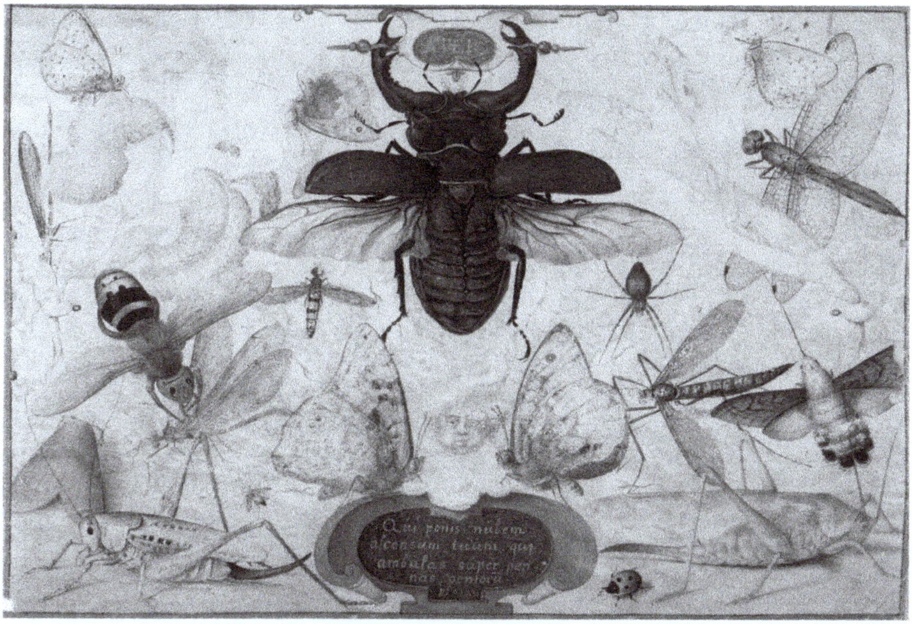

FIGURE 8.5 *Insects and the Head of a Wind God* (*c.* 1590–1600) by Joris Hoefnagel. Pen and brown ink, colored washes, gold paint, on vellum. Wikimedia Commons.

been published, leaving Joris without model insects to draw from. To overcome this, he drew his insect depictions directly from preserved specimens, allowing him to illustrate them true to size (Vignau-Wilberg 2017). His accurate depiction of a mayfly in *Ignis*, the first ever to be published, inspired the son of the director of Leiden University to later publish a book on Ephemeroptera. Joris's detailed illustrations of insects can also be found in Georg Bocksay's *Model Book of Calligraphy and the Constructed Alphabet* (1592–1596) and the *Book of Hours of Philip of Cleves* (1591–1600). Miniatures of ichneumon wasps, damselflies, caterpillars, tiger moths, butterflies, lady beetles, flies, bees, beetles, grasshoppers, and shield bugs drawn by Joris can be found in both books. Joris also included insects in over twenty traditional oil paintings, depicting them in realistic detail. Butterflies, caterpillars, a damselfly, and a variety of hymenopterans can be seen in *Still Life with Flowers, a Snail and Insects for the Artist's Mother*; more lepidopterans, both adults and immatures, a wasp, and a large stag beetle are depicted in *Flower Vase – Friendship Picture for Johannes Radermacher*; craneflies along with the usual butterflies, beetles, and wasps decorate Joris's *The Four Seasons and the Four Ages of Man: Spring, Summer,* and *Autumn* series.

Breaking family tradition, Jacob Hoefnagel was professionally trained as a painter rather than a merchant. Before the age of twenty, Jacob published

Archetypa Studiaque Patris Georgii Hoefnagelii, four volumes of engravings created after models drawn by his father Joris. By the turn of the seventeenth century, Jacob began drawing on his own rather than recreating his father's works (Vignau-Wilberg 2017). Like his father, Jacob depicted insects in both miniatures and oil paintings, but to a lesser extent. Some of Jacob's works that included insects bore both Jacob's and Joris's names, perhaps indicating a collaboration or Jacob's use of his father's insect models as inspiration. Although insects were depicted in fewer of Jacob's paintings, he continued to focus on the portrayal of the natural world, often painting plants and animals. An album of insect depictions, *Diversae insectarum volatilium icones*, including drawings from *Archetypa*, was published by Jacob in 1630. It was around this time that entomology began to establish itself as a new field of study.

WOMEN PIONEERS IN ENTOMOLOGY AND ART

The Age of Enlightenment also is special because it represents a time period where many of the painters were women of entomology (George 2010). Maria Sibylla Merian was a scientist, artist, explorer, and middle-class woman (Valiant 1993). Her works combined detailed observations of the physical forms and life histories of her subjects (Merian and O'Brien-Twohig 2012). A pioneer of both ecology and developmental biology, she forever changed the field of scientific illustration by placing animals in their native environments (Valiant 1993; Kutschera 2017). In typical Enlightenment fashion, Merian traveled abroad to Surinam to document the life cycles of the diverse insects found there, the adults of which she had observed in collections. She was accompanied by her daughter who assisted in her research efforts (Merian and O'Brien-Twohig 2012). Merian's work focused on insect metamorphosis; she studied caterpillars from childhood until her 1717 death (Valiant 1993). Her greater than fifty years of observations became the first long-term zoological endeavor (Valiant 1993). She wrote:

> From my youth onwards I have been concerned with the study of insects, in which I began with silkworms in my native city, Frankfurt am Main; then I observed the far more beautiful butterflies and moths that developed from caterpillars other than silkworms, which led me to collect all the caterpillars I could find in order to study their metamorphosis. I therefore withdrew from society and devoted myself to these investigations; at the same time, I wished to become proficient in the skill of painting in order to paint and describe them from life.
>
> (Merian and O'Brien-Twohig 2012)

Merian fulfilled this wish by training in the art of watercolor, taught to her by her stepfather (Valiant 1993). Her published journal on insects, *Schmetterlinge, Käfer und Andere Insekten*, contains some of her earliest artistic works. She began painting around the age of eight, fully developing her artistic style by the time she was thirteen (Valiant 1993). In her journal she noted details of her insect collections, including dates and locations, and recorded observations of metamorphosis, reproduction, and life cycles of insects, mentioning eggs, frass, and even parasitoid attacks (Valiant 1993). The journal also includes records of hostplant interaction experiments, where she noted that some insects could feed from a variety of plants while others could only tolerate a single species. Even in her earliest records, all of the illustrated insects were painted from life. Merian later published her most famous work, *Metamorphosis Insectorum Surinamensium*, a culmination of her travels and observations in Surinam (Figure 8.6) (Valiant 1993). One

FIGURE 8.6 Caterpillars, Butterflies (*Arsenura armida*) and Flower (*Pallisaden Boom: Erythrina fusca*), Plate 11 from *Metamorphosis Insectorum Surinamensium*, first published 1705. Wikimedia Commons.

of the largest volumes ever published, her book on the insects of Surinam was published by Merian herself. This publication revealed to Europeans the incredible diversity of the tropics (Valiant 1993). Along with her famed illustrations of insect life cycles on their respective host plants, the work also included animal behavior (e.g., gregariousness in caterpillars and the bridge-building of leafcutter ants), remedies and practical uses of plants and animals, ecological interactions, and descriptions of tropical fruits such as pineapple, watermelon, and cassava (Valiant 1993; Merian and O'Brien-Twohig 2012). *Metamorphosis Insectorum Surinamensium* included the life cycles not only of her beloved lepidopterans, but also of beetles, bees, and flies, all on their host plants as well (Merian and O'Brien-Twohig 2012). Sketches of other tropical wildlife, such as spiders, snakes, frogs, and lizards also made their way into the volume. Throughout the text she included both the local common names and Latin names of the plants and animals whenever possible (Merian and O'Brien-Twohig 2012). Throughout Merian's published works is evidence that many important discoveries and ideas in entomology, often attributed to other scientists, were first developed by Merian (Valiant 1993; Kutschera 2017). The metamorphosis of silkworms, for example, published by Malpighi in 1669, was recorded in Merian's journal nearly ten years prior in 1660 (Valiant 1993). Merian wrote about insect eggs in her journal years before Redi published this theory in 1668, which overturned Aristotle's theory of spontaneous generation, an idea that Merian long refuted (Valiant 1993; Kutschera 2017). In the late seventeenth century, Merian also understood that insects come from parents of the same species, a concept that was not fully understood for another hundred years (Valiant 1993). Her descriptions were even organized by classification systems that did not yet exist: she made distinctions between butterflies and moths, grouped a "strange snake," in actuality a legless lizard, with other lizards, and grouped spiders and scorpions, both arachnids (Valiant 1993). Her research garnered her fame in the eighteenth century: she was cited over a hundred times by Carolus Linnaeus, and his student, entomologist Johan Christian Fabricius, cited Merian with even more frequency (Valiant 1993). Moses Harris also cited Merian in *The Aurelian*, an important English work on entomology. It is possible that many species of Lepidoptera should bear the name "Merian"; Linnaeus, Stoll, and Cramer often took over the description of insects studied by Merian, sometimes without ever examining the specimens themselves (Valiant 1993). In addition to her insights on insect reproduction, metamorphosis, and hostplant interactions, Merian also showed that small animals, such as insects, can be portrayed at any size if they remain in proportion to their environment (Valiant 1993).

Still life paintings became very popular during the Age of Enlightenment (Chong and Kloek 1999), and few woman painters were as successful at that time as Rachel Ruysch (1664–1750). Ruysch's style was that of the Rococo

FIGURE 8.7 The painting *Roses, Convolvulus, Poppies, and Other Flowers in an Urn on a Stone Ledge* by Ruysch exhibits different species of butterflies. Wikimedia Commons.

movement (Vigué 2002) and used bright colors in her works, which usually consisted of floral arrangements. The bouquets in her paintings were put together with a haphazard emphasis that created a unique 3D effect and contributed to her fame. Interestingly, many of her works showcase insects, particularly butterflies (Figure 8.7). Her interests in flowers, butterflies, and other animals was likely influenced by her father, who was a professor of botany and anatomy (Cook 2011).

Also a Dutch artist from the Enlightenment, Maria van Oosterwijck (1630–1693) included insects in her still life paintings of flowers. Her work differed from paintings by Rachel Ruysch in that her paintings, rather than simply depicting nature, were heavily influenced by her deeply religious beliefs. She used a Chiaroscuro technique in her paintings to emphasize light and shadow (Vigué 2002), and would include insects on flowers, in particular, she would paint the Red Admiral butterfly (*Vanessa atalanta*).

INSECT ART AND ITS ROLE IN CULTURE

Early in the Enlightenment, seventeenth-century Dutch still lifes, particularly flower still lifes and fruit still lifes, contained an alarming number of often-discounted insect depictions. Dicke (2000) estimated that nearly 80 percent of these Dutch paintings contain insects, from tens to hundreds of insects featured in each work. For example, Jan Olis's *Flower Still Life* (1662) featured the commonly painted Red Admiral, while Jan van Kessel painted over a hundred insects into *Europe* in his *The Continents* series (1664). It is clear that lepidopterans, primarily butterflies, are the most popular insect featured in nonempirical art. The Cabbage White appeared frequently during this time, representing purity, and was often contrasted by the inclusion of the darker *Vanessa* sp., the Red Admiral and Painted Lady butterflies.

Flies also make regular appearances in early Enlightenment still lifes of northern Europe as a symbol of mortality and decay. Dragonflies, damselflies, and those beloved lepidopterans were also used as symbols of the eternal life of the soul—the metamorphosis of moths and butterflies and the change from aquatic to terrestrial environments of odonates symbolize how the soul remains after transition from life to death. Although commonplace in flower and fruit still lifes of northern Europe, those painted in the southern regions of the continent lack such a fervor for insects. The eighteenth century showed a marked decrease in the number of still lifes that featured insects with approximately 40 percent of paintings including entomological themes. The frequency of insects in paintings was similar throughout the eighteenth and nineteenth centuries (Dicke 2000).

Not only were insects the subjects of aesthetic, empirical, and symbolic works of art, but they also played an important role behind the artistic scenes as sources of pigments, especially red pigments derived from hemipterans. The pigment carmine is derived from the hemolymph of the coccoid scale insect, commonly referred to as "cochineal," which feeds on prickly pear cactus. Red pigments and paints derived from the cochineal insect had been used by Indigenous Central and Southern Americans since the pre-Toltec era (*c.* 900 CE), but remained unknown to Europeans until the invasion of conquistadors in the fifteenth and sixteenth centuries (Baranyovits 1978; Dapson 2007). Following its "discovery" by the Spanish, Spain developed a monopoly on cochineal carmine beginning in the early eighteenth century and lasting three hundred years, through the remainder of the Enlightenment (de Jesus Mendez-Gallegos et al. 2003). Demand for this pigment was especially high as the textile industry boomed in the mid-eighteenth century following the advent of mechanized textile production (Baranyovits 1978). Carmine pigments were used by Baroque-era Enlightenment painters such as Rembrandt, who featured Carmine Lake in several of his works (Van de Wetering 2005).

CABINETS OF CURIOSITY

Robert Hooke's research revealed that ordinary, commonplace creatures when viewed under the microscope could be quite as exotic as any from faraway places. To select subjects for study—a flea, a louse, an ant, a fly—was in many ways antithetical to his contemporaries, those who Hooke remarked were focused on Nature's greater and more beautiful creatures. Had he been looking for exotic specimens, Hooke could have accessed the collection of Elias Ashmole (1617–1692), a founding member of the Royal Society who had acquired the Tradescant Ark (a collection assembled by botanist John Tradescant the Elder (1577–1638) and his son John Tradescant the Younger (1608–1662)), as it was known, a vast assortment of botanical, zoological, and ethnographic materials from around the world. This collection was England's first cabinet of curiosities or cabinet of wonders. Ashmole donated his collection to the University of Oxford, which dedicated an entire building to it that is known as the Ashmolean Museum.

Also known as the *Wunderkammern*, they were the forerunners of museums containing biological and mineral specimens, art, both religious and secular, and objects from antiquity and foreign cultures (Figure 8.8). Privately owned, the earliest cabinets appeared in the Renaissance (sixteenth century), but it is generally agreed that by the Baroque era they were highly fashionable. While these collections were often held within a cabinet, the word cabinet originally referred to a room that today we would describe as a study. As such, one could enter a cabinet of curiosities (Mauriès 2002), a completely immersive experience of insatiable collecting, or "A theatre of the broadest scope, containing authentic materials and precise reproductions of the whole universe," as declared by Samuel Quiccheberg in *Inscriptiones vel Tituli Theatri Amplissimi* (1565). In short, a cabinet of curiosities was a museum of everything! Quiccheberg stated that the goal was to "obtain rapidly, easily and safely a true and unique understanding of the world combined with an admirable wisdom" (cited in Mauriès 2002).

Clearly one had to be a person of means to assemble such a cabinet. Royalty, those of noble birth, and wealthy merchants aspired to create the most awe-inspiring, comprehensive collections. Excellence was defined by quantity as well as the rarity or the exotic nature of an object or specimen, such as a saint's relic or a unicorn (narwhal) horn. Cabinets began as a symbol of prestige and later a reflection of one's engagement with the growing interest in science. Regarding the shifting paradigm, cultural historian Celeste Olaquiaga writes,

> Caught between two modes of thinking and living—one traditional and theologically-based; the other modern and materially oriented—as well as registering a deeply changing relation to nature, the history of

FIGURE 8.8 Engraving of a cabinet of curiosity from Ferrante Imperato, *Dell'Historia Naturale* (Naples 1599), but printed in *Venedig* in 1672. Wikimedia Commons.

> these monuments to human inquisitiveness spans almost two centuries of transformation, a period considered “illuminating” for its rational, systematic attempt at understanding how the world works.
>
> (Olaquiaga 2005–2006)

With a growing desire to engage the scientific community as well as amateur naturalists, it was not uncommon to create a printed catalog or book documenting one's collection. As few cabinets remain intact today, these publications are a valuable record of not just the breadth of specimens and objects, but how they were displayed. As time progressed the display was as much a point of interest as the collected objects. This included the furniture and mounts as well as arrangement. Much like installation art today, the visitor was not only to view and appreciate the individual pieces but was intended to have a deep visceral experience.

Paintings and prints document the inclusion of insects as part of many cabinets of wonder, but sadly none of the remaining cabinets on public view contain their original specimens. As insect collections are susceptible to the destructive force of pests and climate, it is easy to imagine that without proper care many specimens simply disintegrated. However, others were dispersed and joined other collections. The contents of a cabinet might contain many thousands of items and thus it was rarely possible to keep them together upon the death of the collector who might have debts to be paid or relatives who did not share the same passion for collecting. An example of one such specimen that traveled through many collections is the oldest known pinned insect, a butterfly, *Pontia daplidice*, commonly known as the Bath White, which dates to 1702. The providence of this specimen is well documented. It was collected by William Vernon (1660–1735) and sold to James Petiver (1665–1718), a founder of British entomology. The butterfly came into the collection of James Charles Dale (1792–1872), an influential entomologist whose estate donated the collection and Dale's archives to the Oxford University Museum of Natural History in 1906, where it resides today along with other historically significant specimens.

Declared historic monuments in 1979, the cabinets of curiosities that once belonged to Joseph Bonnier de La Mosson (1702–1744) in the library of the Museum of Natural History in Paris are situated within a modern library that doesn't allow one the fully immersive experience of a *Wunderkammern*, however, they are breathtaking nonetheless. Bonnier de La Mosson was an aristocrat. The original cabinets, nine in total, were spread throughout his residence. They were devoted to anatomy, chemistry, pharmacy, mechanic and optical devices, fine art, and natural history. The collection was renowned in eighteenth-century Paris for its encyclopedic breadth. Generally agreed to be some of the most beautiful cabinets ever made, sadly the collection was quickly dispersed upon Bonnier de La Mosson's death. What is on view in the library is a relatively small portion of the once great collection and is referred to as the second cabinet.

Two collection catalogs were published, one in 1737 and another in 1744, which were commissioned by Bonnier de La Mosson's family after his death in order to draw interest in a sale as he had passed away quite destitute. Each cabinet section is topped with a different set of antlers, giving an overall impression of five stage sets. The cabinetry exhibits the Rococo style at its most elegant: fluid lines, naturalistic asymmetry, and pastel colours. Celeste Olaquiaga writes that "the Cabinet Bonnier de la Mosson stands as a unique manifestation of the intersection between aesthetics and science." It "has the rare quality of combining the atmospheric mise-en-scène (i.e., the arrangement of scenery and stage properties in a play) of the preceding *Wunderkammern* with the organizational intent of the later cabinets, producing an original blend of system and fantasy" (2005–2006). Cultural historian G.M. van de Roemer (2014) also notes a shifting style, writing,

Earlier collections of the sixteenth and seventeenth centuries usually displayed a heterogeneous assemblage based on the accumulation of individual objects, all juxtaposed in such a way as to emphasize variety, abundance and contrast. In the late seventeenth and early eighteenth centuries, however, a new tendency of ordering began to gain favor. Collectors were focused more on the overall composition of their collection as a totality, paying particular attention to the proportional and aesthetic harmony between the individual objects.

The insects in the Bonnier de La Mosson cabinets are in a geometric arrangement that is both aesthetically pleasing and shows some regard for species, genus, and family, although these considerations are subservient to the overall decorative design. For example, the display on the lower left side is entirely made up of beetles, but surrounding the large scarab specimen (*Goliathus goliathus*) are jewel beetles in the family Buprestidae. Likewise, the display on the right's central focus is a beetle in the family Cerambycidae, which is surrounded by jewel beetles. Looking at the display, the arrangements seem fanciful with only the most casual nod to classification, however, at the time the need to examine, organize, and categorize was a direct response to the scientific age.

FIGURE 8.9 Illustration from the book *Wondertooneel der Nature*—a Cabinet of Curiosities or *Wunderkammern* in Holland. Wikimedia Commons.

In the Netherlands, another man was amassing a similarly impressive collection which he referred to as his "Wondertooneel der Nature" (Wonder Theatre of Nature). The collection of Levinus Vincent (1658–1727) was comprised of eight cabinets (Figure 8.9) all focused on the natural world, the pride of which was "288 boxes of indigenous and exotic insects" (Figure 8.10) (van de Roemer 2014).

Based in Amsterdam, records note that Vincent was a merchant, specifically a damask merchant (a type of woven fabric in which the pattern is reversed on the underside, which was considered a luxury material woven of silk), yet he was also a member of the local painting guild. Thus, it is not surprising that a wealthy businessman with an artistic bent might consider arranging his specimens in the new fashion, creating spiraling patterns of butterflies and moths (Figure 8.11) that reflected the fabrics he sold to wealthy clients. With the help of his wife, Joanna van Breda, the couple meticulously ordered their collection making it a novel destination for visitors to the city who were charged an entrance fee at the door. Scholars, writers, and the curious pondered at what was more enticing, the specimens or their extraordinary presentation of materials.

Today, when viewing illustrations of Vincent's cabinet, one has a tendency to dismiss the collection as lovely, but unscientific—beauty for beauty's sake.

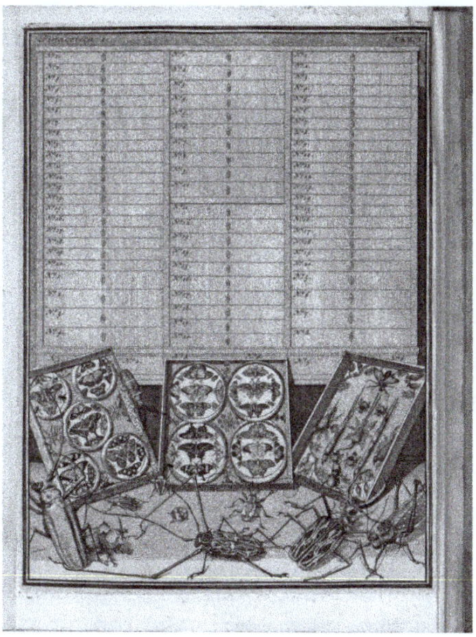

FIGURE 8.10 Illustration from the book *Wondertooneel der Nature*—a Cabinet of Curiosities or *Wunderkammern* in Holland, showing some of the insect collections. Wikimedia Commons.

FIGURE 8.11 Illustration from the book *Wondertooneel der Nature*—a Cabinet of Curiosities or *Wunderkammern* in Holland, showing the spirals made of butterflies. Wikimedia Commons.

By virtue of its name alone, the "Wonder Theatre of Nature," suggests that the visitor was to see a spectacle. There seems little doubt that Vincent's motivation to create the cabinet was partially out of vanity and was highly fashionable at the time for a successful merchant. He boasted that his collection contained all the species noted by the renowned scientific illustrator Maria Sibylla Merian in *Metamorphosis insectorum Surinamensium* (1705) (Spary 2004).

Clearly obsessed with creating an encyclopedic collection illustrating the greatest variation and diversity, Vincent's multiple publications on the cabinet make clear that the study and observation of the natural world was not divorced from religion just yet. He was a pious man who believed that "No-one gazing upon the multiplicity of natural productions, could fail to worship God for His Creation" (Spary 2004). Describing his own part in the undertaking he referred to himself as the "principal maker of order," Thus, the time-consuming precision of ordering the specimens can perhaps be seen as a kind of religious ritual or devotional activity not dissimilar from sand paintings created by Buddhist monks in Tibet. In particular, as a Christian, butterflies would have held significance for Vincent as they were considered a symbol of Christ's resurrection. Undoubtedly, Vincent hoped to renew his visitor's faith in God.

CONCLUSION

The Age of Enlightenment marked a time in human history when our understanding of the world around us shifted from a basis in religion and mythology to one of empiricism, science, and reason. Traditional art depicting or featuring insects during this time was rife with symbolism. Insects were featured as small elements of still life paintings and as the central focus of works by those interested in natural history, such as the Hoefnagel's. Insects in art ranged from fantastical, beastly versions to strikingly realistic illustrations, which without Hooke's advancements in microscopy would not have been possible. The advent of the microscope allowed for insects to be seen for the first time in epic proportions—exemplified by Hooke's massive flea in *Micrographia* (the illustration on the cover of this book). Swammerdam's focus on illustrating insects sparked a massive outpouring during the Enlightenment of illustrated volumes on a variety of entomological themes produced by individuals such as Merian, Harris, Hubner, and Donovan. Their works exemplify the shift from art for the purpose of aesthetics alone to the acceptance of art's importance in conveying empirical information through illustrations created with scientific accuracy in mind. These artists were able to accurately depict thousands of insects by drawing from live specimens; however, the resulting illustrations undoubtedly retain ample amounts of aesthetic beauty. Cabinets of curiosities, fashionable during the Enlightenment, often featured insects, preserved and prepared using methods still practiced by entomologists today, for both empirical and aesthetic purposes, yet again exemplifying the shift of human knowledge towards a more scientific understanding during the Age of Reason.

NOTES

Chapter 7

1. There are countless authors between 1600 and 1820 who use insects in their work. Due to the scope of this project, I have chosen to limit the focus of this chapter and examine only three representative authors whose contributions to literature and language are well recognized.
2. Because this volume covers the span of time from 1600 to 1820, the works of Shakespeare discussed in this chapter are those dated from 1600–1613. For an examination of Shakespeare's earlier works, please see the previous volume in this series, Volume 3, *A Cultural History of Insects in the Renaissance*.
3. While Shakespeare makes a reference to maggots created by the sun in *Hamlet*, in *The Tempest* and *Cymbeline* he writes of "flesh-flies" causing the generation of maggots in decomposing bodies (Patterson 1838: 175–8).
4. Other scholars have provided alternative readings, as is the case with Anthony B. Dawson who explains the line as referring to a "piece of flesh good for kissing" (Shakespeare [1600] 2016: fn. 3, 1791).
5. Several Sicilian folktales and Greek myths tell of women conceiving when the sun's rays touch their bodies (Hankins 1949: 512–13). Edmund Spenser, a contemporary of Shakespeare, described the conception of Belphoebe and Amoret by the sun's rays on the nymph Chrysogone (*The Faerie Queene*, III.vi.7–8).
6. Some critics interpret Mamillius's reference to crickets as alluding to the women standing near them, as Jean E. Howard does in the Norton edition of *The Winter's Tale* (Shakespeare [1609] 2016: 3147).
7. Thomas Hanmer, in his edition of *The Works of Shakespear* [*sic*], in six volumes (London, 1743–1744) began the debate by glossing "shard" as "A SHARD, a tile or broken piece of a tile: thence figuratively a scale or shell upon the back of any Creature. The *Shard-born Beetle* means the Beetle that is born up by wings hard and glazed like a Pot-sheard" (Vol. 5, 327). In 1778, George Steevens reproduced Hanmer's gloss for "shard" in the edition he produced with Samuel Johnson and Isaac Reed, without giving credit to Hanmer where credit was due (*The Plays*

of William Shakspeare [*sic*] 10 vols., London, 1778–1780). Since then, Steevens has been most associated with reading "shard-born beetle" as a reference to the insect's wings. For more on this issue, see Billings (2005).

8. Shakespeare refers to the destructive nature of the worm (cankers, moths, and caterpillars) in a number of pre-1600 plays: *Henry VI Part 2*, *Richard II*, *Romeo and Juliet*, *A Midsummer Night's Dream*, and *Much Ado About Nothing*.
9. Shelley erroneously categorizes the gadfly as a type of beetle.
10. Kellert's five "motivational factors" for humans to be apprehensive toward insects are the following:

    First, many humans are alienated by the vastly different ecological survival strategies, spatially and temporally, of most invertebrates in comparison to humans. Second, the extraordinary "multiplicity" of the invertebrate world seems to threaten the human concern for individual identity and selfhood. Third, invertebrate shapes and forms appear "monstrous" to many people. Fourth, invertebrates are often associated with notions of mindlessness and an absence of feeling – the link between insects, spiders, and madness has been a common metaphor in human discourse and imagination. Fifth, many people appear challenged by the radical "autonomy" of invertebrates from human will and control.

    (1993: 57–8)

11. By Milton's day, such views had changed and he instead praises the silkworm's industriousness: "Worms | That in their green shops weave the smooth-hair'd silk" (*Comus*, ll.715–16).
12. Although 1599 is commonly accepted as the year of the play's composition, *The Chronicle History of Henry the Fifth* was entered into the Register of the Stationers Company on August 14, 1600. Due to its borderline date within the focus of this edition, and the relevancy of the bee analogy here presented to that of Milton's in *Paradise Lost*, I have chosen to include the play in this chapter.
13. Shakespeare mistakenly identifies the queen bee as male, a common assumption at the time.
14. Shakespeare's personal political views are unknown. Because of this, we cannot assume that the idealization of the beehive as an analogy for a perfect monarchy was his own. We can, however, note that such a view is fitting for the character of the Archbishop of Canterbury as an advisor to the king and as someone who is prompting King Henry to invade a foreign country.
15. The insects Milton mentions, and the number of mentions for each one, are the following: bee, 5; fly, 3; locusts, 2; cricket, 1; emmet (ant), 1; lice, 1; hornet, 1. The word "insect" appears three times (Squires 1894).
16. The insects Shelley refers to in his poetry, and their frequency, is as follows: worm, 58; bee, 25; scorpion, 12; moth, 9; spider, 7; glowworm, 6; ephemerid, 5; butterfly, 4; firefly, 4; silkworm, 4; fly, 3; gnat, 3; ant, 2; beetle, 2; centipede, 2; cicada, 2; gadfly, 2; grasshopper, 2; leech, 2; cricket, 1; dragonfly, 1; scarab (beetle), 1 (Jeffrey 1976).
17. Interestingly, Shakespeare notes particularly that drone bees do not suck blood, as Shelley later states in his poem: "Drones suck not eagle's blood, but rob beehives" (*2 Henry VI*, 4.1.116).

BIBLIOGRAPHY

Abbie, A.A. 1960. "Physical Changes in Australian Aborigines Consequent Upon European Contact." *Oceania* 31(2): 140–4.

Adams, E. 1859. "The Vernacular Names of Insects." *Transactions of the Philological Society* 6: 84–96.

Adams, M., Alther, W., Kessler, M., Kluge, M., and Hamburger, M. 2011. "Malaria in the Renaissance: Remedies from European Herbals from the 16th and 17th Century." *Journal of Ethnopharmacology* 133: 278–88.

Adelmann, H.B. 1966. *Marcello Malpighi and the Evolution of Embryology*, 5 vols. (Biography is vol. 1, index in vol. 5). Ithaca: Cornell University Press.

Adler, P.H. and Wills, W. 2003. "Legacy of Death. The History of Arthropod-Borne Human Diseases in South Carolina." *American Entomologist* 49(4): 216–28.

A.J. 1683. "Directions for the Making of Colonies of Bees." In J. Houghton, *A Collection of Letters for the Improvement of Husbandry and Trade*, 374–81. London: J. Houghton.

Aldridge, A.O. 1993. *The Dragon and the Eagle: The Presence of China in the American Enlightenment*. Detroit, MI: Wayne State University Press.

Alexander, J.T. 1974. "Catherine II, Bubonic Plague, and the Problem of Industry in Moscow." *American Historical Review* 79(3): 637–71.

Alfani, G. 2010. "The Effects of Plague on the Distribution of Property: Ivrea, Northern Italy 1630." *Population Studies* 64(1): 61–75.

Allaby, M. 2010. *Ecology: Plants, Animals, and the Environment*. New York: Facts On File, An imprint of Infobase Publishing.

Allcock, H. 1962. *Heraldic Designs: Its Origins, Ancient Forms and Modern Usage*. New York: Tudo Publishing Company.

A. L. O. E. 1874. *Fairy Frisket; or, peeps at insect life*. London: T. Nelson and Sons.

Amsterdam, R., and Schierbeek, A. 1959. *Measuring the Invisible World: The Life and Works of Antoni van Leeuwenhoek*. London: Abelard-Schulman.

Andersen, K.G., Rambaut, A., Lipkin, W.I., Holmes, E.C., and Garry, R.F. 2020. "The Proximal Origin of SARS-CoV-2." *Nature Medicine* 26: 450–2.

Anonymous. 1600. *The Maydes Metamorphosis*. London: Thomas Creede.

Anscombe, G.E.M. 1959. *An Introduction to Wittgenstein's Tractatus*. London: Thoemmes Press.
Applebaum, W. 2000. *Encyclopaedia of the Scientific Revolution*. New York: Garland.
Arber, A. 1942. "Nehemiah Grew (1641–1712) and Marcello Malpighi (1628–1694): An Essay in Comparison." *Isis* 34(1): 7–16.
Armitage, A. 1958. "A Naturalist's Vacation: The London Letters of J. C. Fabricius." *Annals of Science* 14: 116–31.
Armstrong, P.H. 1994. "Human Impact on the Falkland Islands Environment." *Environmentalist* 14(3): 215–31.
Armstrong, P. 2000. *The English Parson-Naturalist: A Companionship Between Science and Religion*. Leominster, UK: Gracewing.
Bacon, F. 1605. *The Advancement of Learning*. London: Cassell & Company, Lmt.
Bacon, F. 1902. *Novum Organum*. New York: P. F. Collier & Sons.
Baldwin, B.T. 1913. "John Locke's Contributions to Education." *The Sewanee Review* 21: 177–87.
Baldwin, S.A. 1986. *John Ray: Essex Naturalist*. Witham, Baldwin's Books.
Baranyovits, F.L.C. 1978. "Cochineal Carmine: An Ancient Dye With a Modern Role." *Endeavor* 2(2): 85–92.
Barber, R. 1992. *Bestiary: Being an English Version of the Bodleian Library, Oxford MS 764*. London: Folio Society.
Bardi, J.S. 2006. *The Calculus Wars: Newton, Leibniz, and the Greatest Mathematical Clash of All Time*. New York: Basic Books.
Barnett, S. 2013. *The Enlightenment and Religion: The Myths of Modernity*. Manchester: Manchester University Press.
Baron, S. 1972. *The Desert Locust*. New York: Charles Scribner's Sons.
Beier, M. 1973. "The Early Naturalists and Anatomists during the Renaissance and Seventeenth Century." In R.F. Smith, T.E. Mittler, and C.N. Smith (eds.), *History of Entomology*, n. 2, 81–94. Palo Alto, CA: Annual Reviews, Inc.
Belloni, L. 1974. "Marcello Malpighi (1628–84), Medicine, Microscopic and Comparative Anatomy, Embryology." *Dictionary of Scientific Biography* 9: 62–6.
Belloni, L. 1975. "Francesco Redi (1626–98), Entomology, Parasitology, Toxicology." *Dictionary of Scientific Biography* 11: 341–3.
Berenbaum, M. 2006. "On the Lives of Insects in Literature." In E.C. Brown (ed.), *Insect Poetics*, 25–45. Minneapolis/London: University of Minnesota Press.
Berenbaum, M.R. 2017. "Pins and Needles." *American Entomologist* 63(2): 67–9.
Bergland, R.L. 2000. *The National Uncanny: Indian Ghosts and American Subjects*. Hanover, NH: University Press of New England.
Bergström, I. 1983. *Dutch Still-Life Painting in the Seventeenth Century*. New York: Hacker Art Books.
Betts, C.J. 1984. *Early Deism in France: From the So-Called "Déístes" of Lyon (1564) to Voltaire's "Lettres philosophiques" (1734)*. The Hague/Boston: M. Nijhoff Publishers.
Bhat, S.A., Bashir, I., and Kamili, A.S. 2009. "Microsporidiosis of Silkworm, *Bombyx mori* L. (Lepidoptera-Bombycidae): A Review." *African Journal of Agricultural Research* 4(13): 1519–23.
Billings, T.J. 2005. "Squashing the 'Shard-Borne Beetle' Crux: A Hard Case With a Few Pat Readings." *Shakespeare Quarterly* 56(4): 434–47.
Birch, T. 1756. *The History of the Royal Society of London for Improving of Natural Knowledge, from Its First Rise*. London: Royal Society, 1756–1757.

Blair, A. 2017. "Humanism and Printing in the Work of Conrad Gessner." *Renaissance Quarterly* 70(1): 1–43.
Blake, W. 1794. *Songs of Innocence and of Experience*. Self published.
Blunt, W. 1971. *The Compleat Naturalist: A Life of Linnaeus*. London: William Collins and Sons.
Boatin, B.A. and Richards, Jr., F.O. 2006. "Control of Onchocerciasis." *Advances in Parasitology* 61: 349–94.
Bodenheimer, F.S. 1959. *The History of Biology: An Introduction*. London: William Dawson and Sons.
Bonnet, C. 1764. *Contemplation de la Nature*. Amsterdam: M.M. Rey.
Booth, C. 2005. "The Rod of Aesculapios: John Haygarth (1740–1827) and Perkins' Metallic Tractors." *Journal of Medical Biography* 13(3): 155–61.
Bown, N. 2001. *Fairies in Nineteenth-Century Art and Literature*. Cambridge: Cambridge University Press.
Bradley, T.J. 1985. "The Excretory System: Structure and Physiology." In G.A. Kerkut and L.I. Gilbert (eds.), *Comprehensive Insect Physiology, Biochemistry and Pharmacology*, vol. 4, 421–65. New York: Pergamon Press.
Bridoul, F.T. 1687. *Upon the miraculous respects and acknowledgments, which beasts, birds, and insects, upon several occasions, have rendered to the holy sacrament of the altar*. London: Randall Taylor.
Briggs, K. 1976. *An Encyclopedia of Fairies*. New York: Pantheon Books.
Brittan, F. 2011. "On Microscopic Hearing: Fairy Magic, Natural Science and the Scherzo Fantastique." *Journal of the American Musicological Society* 64(3): 527–600.
Brodhead, M.J. 1993. "Thomas Say: New World Naturalist by Partricia Tyson Stroud." *Journal of the Early Republic* 13(3): 428–30.
Brown, E.C. 2006a. "Introduction." In E.C. Brown (ed.), *Insect Poetics*, ix–2. Minneapolis/London: University of Minnesota Press.
Brown, E.C. 2006b. "Performing Insects in Shakespeare's *Coriolanus*." In E.C. Brown (ed.), *Insect Poetics*, 29–57. Minneapolis/London: University of Minnesota Press.
Brown, K. 1993. "Numismatic Entomology." *Cultural Entomology Digest* 1(1): 13–14.
Brown, R. 1994. *Great Masters of Beekeeping*. Somerset, England: Bee Books New & Old.
Brown, S.W. 1973. "Genetics—the Long Story." In R.F. Smith, T.E. Mittler, and C.N. Smith (eds.), *History of Entomology*, n. 2, 4414. Palo Alto, CA: Annual Reviews, Inc.
Browne, W. 1845. *Brittania's Pastorals*. London: H. G. Clarke and Co.
Bryk, F. 1952. "Charles de Geer, 1720–778." In S. Lindroth (ed.), *Swedish Men of Science, 1650–1950*, 113–21. Trans. B. Anderson. Stockholm: Swedish Institute.
Bryson, N. 1990. *Looking at the Overlooked: Four Essays on Still Life Painting*. Cambridge, MA: Harvard University Press.
Buckland, P.C. and Sadler, J.P. 1989. "A Biogeography of the Human Flea, *Pulex irritans* L. (Siphonaptera: Pulicidae)." *Journal of Biogeography* 16(2): 115–20.
Budden, F.H. 1957. "Natural History of Onchocerciasis." *British Journal of Ophthalmology* 41: 214–26.
Budgen, L.M. 1851. *Episodes of Insect Life*, 3 vols. Clinton-Hall, NY: J. S. Redfield.
Burkhardt Jr., R.W. 1973. "Pierre-André Latreille (1762–1833), Entomology, Zoology." *Dictionary of Scientific Biography* 8: 48–9.

Burkhardt Jr., R.W. 2013. "Lamarck, Evolution, and the Inheritance of Acquired Characters." *Genetics* 194(4): 793–805.

Burns, J.N., Acuna-Soto, R., and Stahle, D.W. 2014. "Drought and Epidemic Typhus, Central Mexico, 1655–1918." *Emerging Infectious Diseases* 20(3): 442–7.

Butler, C. [1609] 1623. *The Feminine Monarchie: Or the Historie of Bees*. London: John Haviland.

Butler, C. 1634. *The Feminine Monarchie or the Histiori of Bees*. Oxford: William Turner.

Butler, C. 1704. *The Feminine Monarchie or the History of Bees*. London: A. Baldwin.

Butlin, M. 1981. *The Paintings and Drawings of William Blake*, vol. 1. New Haven, CT: Yale University Press.

Byrne, J.M. 1997. *Religion and the Enlightenment: From Descartes to Kant*. Louisville, KY: Westminster John Knox Press.

Calman, W.T. 1939. "Caroli Linnaei Systema Naturae." *Nature* 144: 269.

Cambefort, Y. 2006. "A Sacred Insect on the Margins: Emblematic Beetles in the Renaissance." In E.C. Brown (ed.), *Insect Poetics*, 39–49. Minneapolis/London: University of Minnesota Press.

Cameron, J.W.M. 1973. "Insect Pathology." In R.F. Smith, T.E. Mittler, and C.N. Smith (eds.), *History of Entomology*, n. 2, 28628. Palo Alto, CA.: Annual Reviews, Inc.

Capinera, J.L. 2008. *Encyclopedia of Entomology*. Dordrecht: Springer Science.

Carlson, D.B., Dale, P.E., Kurucz, N., Dwyer, P.G., Knight, J.M., Whelan, P.I., and Richards, D.D. 2019. "Mosquito Control and Coastal Development: How They Have Coexisted and Matured in Florida and Australia." *Journal of the American Mosquito Control Association* 35: 123–34.

Castellani, C. 1970. "Ulisse Aldrovandi (1522–1605), Natural Sciences." *Dictionary of Scientific Biography* 1: 108–10.

Castillo, M. 2013. "The Scientific Method: A Need for Something Better?" *American Journal of Neuroradiology* 34(9): 1669–71.

Chalmers-Hunt, J.M. 1994. "Entomological Bygones, or Historical Entomological Collecting Equipment and Associated Memorabilla." *Archives of Natural History* 21: 357–78.

Chambers, A.B. 1996. "The Fly in Donne's 'Canonization'." *The Journal of English and Germanic Philology* 65(2): 252–9.

Chastel, A. 1984. *Musca depicta*. Milan, Italy: Franco Maria Ricci.

Chastel, A. 1986. "Iconology of the Fly." *FMR* 4(19): 62–81.

Cherry, R. 2008 "Mythology and Insects." In J.L. Capinera, *Encyclopedia of Entomology*, 2540–3. Dordrecht: Springer.

Cherry, R. 2019. "Lepidoptera in the Mythology of Native Americans." Orkin. Available online: https://www.insects.orkin.com/ced/issue-4/native-american-mythology/ (accessed May 17, 2023).

Cheshire, F.R. 1886. *Bees & Bee-Keeping*, vol. 1. London: L. Upcott Gill.

Chhem, R.K. 2013. "Monsoons, Mosquitoes and Malaria: Alfred Russel Wallace in the Malay Archipelago." *Jati* 18: 247–54.

Chong, A. and Kloek, W. 1999. *Still Life Paintings from the Netherlands 1550–1720*. De Broeren, Zwolle: Waanders Publishers.

Chorley, R.J., and R.P. Beckinsale. 1991. *The History of the Study of Landforms Or the Development of Geomorphology*. Abingdon: Routledge.

Clark, J.F.M. 2004a. "William Kirby (1759–1850)." *Dictionary of Nineteenth-Century British Scientists* 3: 1147–8.
Clark, J.F.M. 2004b. "William Spence (1782–1860)." *Dictionary of Nineteenth-Century British Scientists* 4: 1883–4.
Clark, J.F.M. 2004c. "William Spence (178 –1860), Political Economist and Entomologist." *Oxford Dictionary of National Biography* 51: 813–14.
Clark, J.F.M. 2009. *Bugs and Victorians*. New Haven, CT: Yale University Press.
Clarke, C. and Clarke, M.C. 1886. *Cassel's Illustrated Shakespeare*, vol. 1. Comedies. London: Cassell, Petter, & Galpin.
Clerck, C. 1757. *Svenska Spindlar uti sina hufvud-slågter indelte samt under några och sextio särskildte arter beskrefne och med illuminerade figurer uplyste/Aranei Svecici, descriptionibus et figuris æneis illustrati, ad genera subalterna redacti, speciebus ultra LX determinati*. Stockholm: Laurentius Salvius
Coakley, S. 2012. *Faith, Rationality and the Passions*. Hoboken, NJ: Blackwell Publishing Ltd.
Cobb, M. 2002. "Exorcizing the Animal Spirits: John Swammerdam on Nerve Function." *Nature Reviews* 3: 395–400.
Cobb, M. 2006. *Generation: The Seventeenth-Century Scientists Who Unraveled the Secrets of Sex, Life, and Growth*. London: Bloomsbury Publishing.
Cockburn, T.A. 1971. "Infectious Disease in Ancient Populations." *Current Anthropology* 12(1): 45–62.
Cockerell, T. 1922. "Dru Drury, an Eighteenth Century Entomologist." *The Scientific Monthly* 14(1): 67–82.
Cohen of Birkenhead. 1971. "Erasmus Darwin (1731–1802), Medicine, Scientific Poetry, Botany, Technology." *Dictionary of Scientific Biography* 3: 577–81.
Cohen, B.I. 1965. "Reviewed Work: The Scientific Renaissance, 1450–1630, Marie Boas." *Isis* 56(2): 240–2.
Cohen, B.I. 1976. "The Eighteenth-Century Origins of the Concept of Scientific Revolution." *Journal of the History of Ideas* 37(2): 257–88.
Cole, F.J. 1944. *A History of Comparative Anatomy*. London: Macmillan.
Coleman, D. 2006. "Entertaining Entomology: Insects and Insect Performers in the Eighteenth Century." *Eighteenth-Century Life* 30(3): 107–34.
Connor, S. 2007. "Flysight." *Cabinet* 25: 78–84.
Cook, H.J. 2011. "L. Kooijmans, Death Defied: The Anatomy Lessons of Frederick Ruysch." *BMGN – Low Countries Historical Review* 126(4): 126–7.
Cooper, M. and Hunter, M. 2006. *Robert Hooke: Tercentennial Studies*. Aldershot, Ashgate.
Coutts, N. 2006. "Portraits of the Nonhuman: Visualizations of the Malevolent Insect." In E.C. Brown (ed.), *Insect Poetics*, 298–318. Minneapolis/London: University of Minnesota Press.
Cowan, F. 1865. *Curious facts in the history of insects including spiders and scorpions*. Philadelphia, PA: J.B. Lippincott & Co.
Cox, F.E.G. 2002. "History of Human Parasitology." *Clinical Microbiology Reviews* 15(4): 595–612.
Crane, E. 1999. *The World History of Beekeeping and Honey Hunting*. London: Duckworth.
Crichton, M. 1990. *Jurassic Park*. New York: Ballantine Books.
Crosby, A.W. 1976. "Virgin Soil Epidemics as a Factor in the Aboriginal Depopulation in America." *William and Mary Quarterly* 33(2): 289–99.

Crump, A., Morel, C.M., and Omura, S. 2012. "The Onchocerciasis Chronicle: From the Beginning to the End?" *Trends in Parasitology* 28(7): 280–8.
Cunningham, A. and Grell, O.P. 2010. *The Four Horsemen of the Apocalypse: Religion, War, Famine and Death in Reformation Europe*. Cambridge: Cambridge University Press.
Cussans, J.E. 1970. *Handbook of Heraldry*, 4th edn. reprint of 1893 edition. Detroit, MI: Gale Research Company.
Cutler, S.J. 2006. "Possibilities for Relapsing Fever Reemergence." *Emerging Infectious Diseases* 12(3): 369–74.
d'Aguilar, J. 2006. *Histoire de l'Entomologie*. Paris: Delachaux & Niestlé.
Damkaer, D.M. 2002. *Johan Christian Fabricius. The Copepodologist's Cabinet: A Biographical and Bibliographical History*, vol. 240 of Memoirs of the American Philosophical Society.
Dapson, R. 2007. "The History, Chemistry and Modes of Action of Carmine and Related Dyes." *Biotechnic & Histochemistry* 82(4–5): 173–87.
Darwin, E. 1791. *The botanic garden. A poem, in two parts. Part I. Containing the economy of vegetation. Part II. The loves of the plants: With philosophical notes*. London.
Darwin, E. 1794. *Zoonomia; or the Laws of Organic Life*. Cambridge: Cambridge University Press.
Daston, L. 1998. "The Nature of Nature in Early Modern Europe." *Configurations* 6(2): 149–72.
Daumas, M. 1964. *A History of Technology and Invention*, vol. 2: *The First Stages of Mechanization*. Trans. Eileen B. Hennessy. New York: Crown.
Davidson, G.R., Hill, C.A., and Wolgemuth, K. 2017. "The Need for a Paradigm Shift in Science Advocacy." *GSA Today* 27(7): 58–9.
Davidson, I. 2012. *Voltaire: A Life*. New York: Pegasus Books.
Davis, G.K. 2012. "Cyclical Parthenogenesis and Viviparity in Aphids as Evolutionary Novelties." *Journal of Experimental Zoology* 318B: 448–59.
de Blécourt, W. 2018. "The Magic of the Printed Word." In *Tales of Magic, Tales in Print*. Manchester, England: Manchester University Press.
de Graft-johnson, J.C. 1957. "African Empires of the Past." *Présence Africaine* 13: 58–64.
de Jesus Mendez-Gallegos, S., Panzavolta, T., and Tiberi, R. 2003. "Carmine Cochineal *Dactylopius coccus* Costa (Rhynchota: Dactylopiidae): Significance, Production and Use." *Advances in Horticultural Sciences* 17(3): 165–71.
DeJean, J.E. 1997. *Ancients Against Moderns: Culture Wars and the Making of a Fin de Siècle*. Chicago, IL: University of Chicago Press.
Delbourgo, J. 2006. *A Most Amazing Scene of Wonders: Electricity and Enlightenment in Early America*. Cambridge, MA: Harvard University Press.
Dennys, R. 1982. *Heraldry and the Heralds*. London: Jonathan Cape.
Derrick, E.H. 1937. "'Q' Fever, a New Fever Entity: Clinical Features, Diagnosis and Laboratory Investigation." *Medical Journal of Australia* 2(8): 281–99.
Desquesnes, M., Holzmuller, P., Lai, D.H., Dargantes, A., Lun, Z.R., and Jittaplapong, S. 2013. "*Trypanosoma evansi* and Surra: A Review and Perspectives on Origin, History, Distribution, Taxonomy, Morphology, Hosts, and Pathogenic Effects." *BioMed Research International*. Available online: http://dx.doi.org/10.1155/2013/194176 (accessed May 17, 2023).
Dicke, M. 2000. "Insects in Western Art." *American Entomologist* 46(4): 228–37.

Dobell, C. 1932. *Antony van Leeuwenhoek and His "Little Animals": being some Account of the Father of Protozoology and Bacteriology and His Multifarious Discoveries in these Disciplines*. London: Constable. New York: Russell & Russel, 1958.

Dobie, M. 2009. "The Enlightenment at War." *PMLA* 124: 1851–4.

Dobson, J. 1969. *John Hunter*. Edinburgh and London: E. & S. Livingstone Ltd.

Dorson, R.M. 1993. *American Folklore*. Chicago, IL: University of Chicago Press.

Dower, J.W. 1971. *The Elements of Japanese Design: A Handbook of Family Crests, Heraldry and Symbolism*. New York: Walker/Weatherhill.

Drake, E.T. 1996. *Restless Genius: Robert Hooke and His Earthly Thoughts*. Oxford: Oxford University Press.

Drake, S. 1971. "Federico Cesi (1585–1630), Botany, Scientific Organization." *Dictionary of Scientific Biography* 3: 179–80.

Drancourt, M., Tran-Hung, L., Courtin, J., de Lumley, H., and Raoult, D. 2005. "*Bartonella quintana* in a 4000-Year-Old Human Tooth." *Journal of Infectious Diseases* 191(4): 607–11.

Drancourt, M., Signoli, M., Dang, L.V., Bizot, B., Roux, V., Tzortzis, S., and Raoult, D. 2007. "*Yersinia pestis* Orientalis in Remains of Ancient Plague Patients." *Emerging Infectious Diseases* 13(2): 332–3.

Dreisbach, D. 2003. *Thomas Jefferson and the Wall of Separation Between Church and State*. New York: NYU Press.

Dunkel, A.R., Zijlstra, J.S., and Groves, C.P. 2011–2012. "Giant Rabbits, Marmosets, and British Comedies: Etymology of Lemur Names, Part 1." *Lemur News* 16: 64–70.

Durant, W. and Durant, A. 1967. *The Story of Civilization, Rousseau and Revolution*. New York: MJF Books.

Eddy, P.F. 1931a. "Insects in English Poetry (Part 1)." *The Scientific Monthly* 33(1): 53–73.

Eddy, P.F. 1931b. "Insects in English Poetry (Part 2)." *The Scientific Monthly* 33(2): 148–63.

Editors, C.R. 2018. *Blaise Pascal: The Life and Legacy of the Legendary French Mathematician and Theologian*. CreateSpace Independent Publishing Platform.

Edwards Jr., C.H. 1984. "The Calculus According to Newton and Leibniz." In D.M. Campbell and J.C. Higgins (eds.), *Mathematics: People, Problems, Results*, vol. 2, 104–11. Belmont, CA: Wadsworth Inc.

Edwards, K.L. 1999. *Milton and the Natural World: Science and Poetry in Paradise Lost*. Cambridge: Cambridge University Press.

Egerton, F.N. 2005a. "History of Ecological Sciences, Part 16: Robert Hooke and the Royal Society of London." *Ecological Society of America Bulletin* 86: 93–101.

Egerton, F.N. 2005b. "History of Ecological Sciences, Part 17: Invertebrate Zoology and Parasitology during the 1600s." *Ecological Society of America Bulletin* 86: 135–8.

Egerton, F.N. 2005c. "History of Ecological Sciences, Part 18: John Ray and His Associates Francis Willughby and William Derham." *Ecological Society of America Bulletin* 86: 301–13.

Egerton, F.N. 2006a. "History of Ecological Science, Part 19: Leeuwenhoek's Microscopic Natural History." *Ecological Society of America Bulletin* 87: 47–58.

Egerton, F.N. 2006b. "History of Ecological Sciences, Part 21: Réaumur and His History of Insects." *Ecological Society of America Bulletin* 87: 212–24.

Egerton, F.N. 2007. "History of Ecological Sciences Part 23: Linnaeus and the Economy of Nature." *Ecological Society of America Bulletin* 88: 72–88.
Egerton, F.N. 2008. "History of Ecological Sciences, Part 30: Invertebrate Zoology and Parasitology during the 1700s." *Ecological Society of America Bulletin* 89: 407–33.
Egerton, F.N. 2012. *Roots of Ecology: Antiquity to Haeckel*. Berkeley: University of California Press.
Eilberg-Schwartz, H. 1989. "Witches of the West: Neopaganism and Goddess Worship as Enlightenment Religions." *Journal of Feminist Studies in Religion* 5(1): 77–95.
Eliot, George [1871] (2004) *Middlemarch*. Ed. G. Maertz. Note by editor of 2004 edition, G. Maertz. Broadview Press.
Emmanouil A., Bechah Y., and Raoult, D. 2016. "The History of Epidemic Typhus." *Microbiology Spectrum* 4(4): PoH-0010-2015.
Enenkel, K.A.E. and Smith, M.S. 2007. *Early Modern Zoology: The Construction of Animals in Science, Literature and the Visual Arts*. Leiden: Brill.
Epstein, J. 2008. The End of the Gutenberg Era. *Library Trends* 57(1): 8–16.
Escalante, A.A., Cornejo, O.E., Freeland, D.E., Poe, A.C., Durrego, E., Collins, W.E., and Lal, A.A. 2005. "A Monkey's Tale: The Origin of *Plasmodium vivax* as a Human Malaria Parasite." *PNAS* 102(6): 1980–5.
Essig, E.O. 1931. *A History of Entomology*. New York: The Macmillan Co.
Farrell, W.J. 1985. "The Role of Mandeville's Bee Analogy in 'The Grumbling Hive'." *Studies in English Literature, 1500–1900* 25(3): 511–27.
Favaro, A. 1900. *Le Opere di Galileo Galilei*. Florence, Barbera: Edizione Nazionale.
Fessenden, T.G. 1804. *Terrible Tractoration!! A Poetical Petition Against Galvanising Trumpery, and the Perkinistic Institution: In Four Cantos: Most Respectfully Addressed to the Royal College of Physicians, by Christopher Caustic, M.D. L.L.D. A.S.S. Fellow of the Royal College of Physicians, Aberdeen, and Honorary Member of No Less Than Nineteen Very Learned Societies*. First American edition from the second London edition. New York: Samuel Stansbury.
Findlen, P. 2000a. "Ulisse Aldrovandi (1522–1605)." In W. Applebaum (ed.), *Encyclopedia of the Scientific Revolution from Copernicus to Newton*, 29–35. New York: Garland.
Findlen, P. 2000b. "Francesco Redi (1626-97)." In W. Applebaum (ed.), *Encyclopedia of the Scientific Revolution from Copernicus to Newton*, 554–6. New York: Garland.
Findlen, P. 2000c. "Controlling the Experiment: Rhetoric Court Patronage and the Experimental Method of Francesco Redi." *History of Science* 31: 35–64.
Findlen, P. 2004. *Athanasius Kircher: The Last Man Who Knew Everything*. Abingdon: Routledge.
Ford, B.J. 1993. *Images of Science: A History of Scientific Illustration*. New York: Oxford University Press.
Forsberg, L. 2015. "Nature's Invisibilia: The Victorian Microscope and the Miniature Fairy." *Victorian Studies* 57(4): 638–66.
Fournier, M. 1990. "The Book of Nature: Jan Swammerdam's Microscopical Investigations." *Tractrix* 2: 1–24.
Fox-Davies, A.C. 1968. *The Art of Heraldry: An Encyclopaedia of Armory*. New York: Benjamin Blom.
Frängsmyr, T. 1983. *Linnaeus: The Man and His Work*. Berkeley: University of California Press.

Fraser, H.M. 1951. *Anton Janscha on the Swarming of Bees: A Critical Survey*. Apis Club, Foxton, United Kingdom.
Fraser, H.M. 1958. *History of Beekeeping in Britain*. London: Bee Research Association.
Freegard, J. 2011. *Kingdom Animalia: The Escapades of Linnaeus*. Auckland: Auckland University Press.
French Académie des Sciences. 2019. Available online: https://www.academie-sciences.fr/en/ (accessed September 28, 2019).
Furuse, Y., Suzuki, A., and Oshitai, H. 2010. "Origin of Measles Virus: Divergence From Rinderpest Virus Between the 11th and 12th Centuries." *Virology Journal* 7: 52.
Gage, A.T. and Stearn, W.T. 1988. *A Bicentenary History of the Linnean Society of London*. Linnean Society of London.
Gage, K.L. and Kosoy, M.Y. 2005. "Natural History of Plague: Perspectives From More Than a Century of Research." *Annual Review of Entomology* 50: 505–28.
Garrison, F.H. 1926. "The Destruction of the Quarantine Station on Staten Island in 1858." *Bulletin of the New York Academy of Medicine* 2: 1–5.
Gaunt, P. 2014. *The English Civil War: A Military History*. London: I.B. Tauris and Co., Ltd.
Gay, P. 1969. *The Enlightenment: An Interpretation*, vol. 2, *The Science of Freedom*. New York: Alfred A. Knopf.
Gedde, J. 1675. *A New Discovery of an Excellent Method of Bee-Houses and Colonies*. London: D. Newman.
George, S. 2010. "Animated Beings: Enlightenment Entomology for Girls." *Journal for Eighteenth-Century Studies* 33(4): 487–505.
Gibson, S. 2017. "The Careering Naturalists: Creating Career Paths in Natural History, 1790–1830." *Archives of Natural History* 44(2): 195–214.
Gilbert, P. 2004. "Edward Donovan (1768–1837), Natural Historian and Artist." *Oxford Dictionary of National Biography* 16: 552.
Gilbert, M.T.P., Cuccui, J., White, W., Lynnerup, N., Titball, R.W., Cooper, A., and Prentice, M.B. 2004. "Absence of *Yersinia pestis*-Specific DNA in Human Teeth From Five European Excavations of Putative Plague Victims." *Microbiology* 150: 341–54.
Gillis, C.L. 1993. "Legendary Pest Remedies." *EPA Journal* 19(4): 44.
Gingerich, O. 1992. "Kepler, Galilei, and the Harmony of the World." In V. Coelho (ed.), *Music and Science in the Age of Galileo*, 45–63. Heidelberg: Springer.
Greenberg, S.J. 1997. "The 'Dreadful Visitation': Public Health and Public Awareness in Seventeenth-Century London." *Bulletin of the Medical Library Association* 85(4): 391–401.
Greenblatt, S. [1980] 2005. *Renaissance Self-Fashioning: From Moore to Shakespeare*. Chicago/London: University of Chicago Press.
Grimaldi, D.A. and Engel, M.S. 2005. *Diversity and Evolution. Evolution of the Insects*, vol. 1 of Cambridge Evolution Series. Cambridge: Cambridge University Press.
Grimm, J., Grimm, W., and Mondschein, K. 2011. *Grimms Fairy Tales*. San Diego: Canterbury Classics.
Guiley, R.E. 2009. *The Encyclopedia of Demons and Demonology*. New York: Checkmark Books.
Gustavsen, K., Hopkins, A., and Sauerbrey, M. 2011. "Onchocerciasis in the Americas: From Arrival to (Near) Elimination." *Parasites and Vectors* 4: 205.

Hajdu, S.I. 2002. "A Note From History: Elephantiasis." *Annals of Clinical and Laboratory Science* 32(2): 207–9.
Hall, R.A. 2002. *Philosophers at War: The Quarrel between Newton and Leibniz.* Cambridge: Cambridge University Press.
Hall, T.S. 1951. "Source Book in Animal Biology, 295–298 Quoted in English Translation the Relevant Text on this Experiment from Bonnet. 1745." *Traité d'Insectologie*, 2 vols. Paris, vol. 1.
Halstead, S.B. 2015. "Reappearance of Chikungunya, Formerly Called Dengue, in the Americas." *Emerging Infectious Diseases* 21(4): 557–561.
Hammon, W.McD. and Reeves, W.C. 1952. "California Encephalitis Virus—A Newly Described Agent." *California Medicine* 77(5): 303–9.
Hampson, N. 1968. *The Enlightenment.* Baltimore, MD: Penguin Books.
Hand, W.D. 1964. *Popular Beliefs and Superstitions from North Carolina*, vol. 7 of The Frank C. Brown Collection of North Carolina Folklore. Durham, NC: Duke University Press.
Hankins, J.E. 1949. "Hamlet's 'God Kissing Carrion': A Theory of the Generation of Life." *PMLA* 64(3): 507–16.
Hannam, J. 2009. *God's Philosophers: How the Medieval World Laid the Foundation of Modern Science.* London: Icon Books Ltd.
Harley, D., Sleigh, A., and Ritchie, S. 2001. "Ross River Virus Transmission, Infection, and Disease: A Cross-Disciplinary Review." *Clinical Microbiology Reviews* 14(4): 909–32.
Harrison, T.P. 1971. "Shakespeare's Glowworms." *Shakespeare Quarterly* 22(4): 395–6.
Hartlib, S. 1655. *The Reformed Commonwealth of Bees.* London: Giles Calvert.
Harvey, W. 1628. *Exercitatio anatomica de motu cordis et sanguinis in animalibus.* Available online: http://special.lib.gla.ac.uk/exhibns/month/june2007.html (accessed May 17, 2023).
Harvey, W. 1651. *Exercitationes de Generatione Animalium.* London.
Hedges Jr., T.R. 2007. "Charles Bonnet, His Life, and His Syndrome." *Survey of Opthamology* 52(1): 111–14.
Hemming, F. 1937. *Hübner: A Bibliographical and Systematic Account of the Entomological Works of Jacob Hübner.* London: Royal Entomological Society of London.
Heniger, J. 1973. "Antoni van Leeuwenhoek (1632–1723), Natural Sciences, Microscopy." *Dictionary of Scientific Biography* 8: 126–30.
Hetzel, P.J. 1977. *Public and Private Life of Animals.* London: Paddington Press.
Hogue, C.L. 1975. "The Insect in Human Symbolism." *Terra* 13(3): 3–9.
Hogue, C. 1979. "The Bugfolk." *Terra* 17: 36–8.
Hollingsworth, C. 2001. *Poetics of the Hive: The Insect Metaphor in Literature.* Iowa City: University of Iowa Press.
Honigsbaum, M. 2003. *The Fever Trail.* New York: Picador.
Hooke, R. 1665. *Micrographia: or, some physiological descriptions of minute bodies made by magnifying glasses. With observations and inquiries thereupon.* London: J. Martyn and J. Allestry.
Hooke, R. 1674. *An Attempt to Prove the Motion of the Earth from Ob-servations.* Reprinted in R.T. Gunther, *Early Science in Oxford*, vol. 3, 1–28. Oxford: Printed for the author (1931).
Huang, S., Molaei, G., and Andreadis, T.G. 2011. "Reexamination of *Culex pipiens* Hybridization Zone in the Eastern United States by Ribosomal DNA-Based Single

Nucleotide Polymorphism Markers." *American Journal of Tropical Medicine and Hygiene* 85(3) 434–41.

Hughes, A.M.D. 1918. "Shelley and Nature." *The North American Review* 208(753): 287–95.

Humphreys, M. 2006. "A Stranger to our Camps: Typhus in American History." *Bulletin of Historical Medicine* 80(2): 269–90.

Hunter, J. 1792. "Observations on Bees." *Philosophical Transactions of the Royal Society* 82: 128–95.

Hutchinson, R.A. and Lindsay, S.W. 2006. "Malaria and Deaths in the English Marshes." *The Lancet* 367: 1947–51.

IBRA (International Bee Research Association). 1979. *British Bee Books: A Bibliography 1500–1976*. London: IBRA.

Impelluso, L. 2004. *Nature and Its Symbols*. Los Angeles, CA: J. Paul Getty Museum.

Impey, O. and MacGregor, A. 1985. *The Origins of Museum: The Cabinet of Curiosities in Sixteenth- and Seventeenth-Century Europe*. Oxford: Clarendon Press.

Jacobs, N.A. 2015. "John Milton's Beehive, from Polemic to Epic." *Studies in Philology* 112(4): 798–816.

Jardine, L. 2003. *The Curious Life of Robert Hooke: The Man Who Measured London.* New York: Harper Collins.

Jeffrey, L.N. 1972/1973. "'Beasts of the Woods and Wildernesses' in the Poetry of Shelley." *Keats-Shelley Journal* 21/22: 64–82.

Jeffrey, L.N. 1976. "Shelley's 'Plumèd Insects Swift and Free'." *Keats-Shelley Journal* 25: 103–21.

Jervis, M.A. 2013. "Robert Hooke's *Micrographia*: An Entomologist's Perspective." *Journal of Natural History* 47. Available online: https://doi.org/10.1080/00222933.2013.780270 (accessed May 17, 2023).

Jervis, M.A. 2014. "A Zoologist's Perspective on Robert Hooke's *Micrographia* (1665) Studies of Marine and Terrestrial Invertebrates, and His Contemplations on Invertebrate "Generation" and Mutability." *Journal of Natural History* 48. Available online: https://doi.org/10.1080/00222933.2013.856492 (accessed May 17, 2023).

Johnston, P. 2011. "A Theatre of Insects, or How Nature Lost Her Morality." *AA Files* 63: 36–45.

Johnston, R.K. and Smith, J.W. 2003. *Life Is Not Work, Work Is Not Life: Simple Reminders for Finding Balance in a 24-7 World*. Tulsa, OK: Council Oak Books.

Kacane, I. 2012. "Insects in the 19th Century British Culture: Pre-Raphaelitism and Aestheticism." *Nature and Culture* 4.

Kamen-Kaye, D. 1979. "A Bug and a Bonfire." *Journal of Ethnopharmacology* 1(2): 103–10.

Keightley, T. 1968. *The Fairy Mythology, Illustrative of the Romance and Superstition of Various Countries*. New York: Haskell House Publishers.

Keiser, I. 1966. "Insects and Related Arthropods in Heraldry." *Bulletin of the Entomological Society of America* 12: 314–18.

Keller, A.G. 1976. "Francesco Stelluti (1577–1652), Microscopy, Scientific Organization." *Dictionary of Scientific Biography* 13: 29–30.

Kellert, S. 1993. "The Biological Basis for Human Values of Nature." In S.R. Kellert and E.O. Wilson (eds.), *The Biophilia Hypothesis*, 42–69. Washington, DC: Island Press.

Kellert, S.R. 1993. "Values and Perceptions of Invertebrates." *Conservation Biology* 7(4): 845–55.
King, L.J. 1975. "Christian Konrad Sprengel (1750–1816), Botany." *Dictionary of Scientific Biography* 12: 587–91.
King-Hele, D. 1963. *Erasmus Darwin*. New York: Charles Scribner's Sons.
King-Hele, D. 1999. *Erasmus Darwin: A Life of Unequaled Achievement*. London: Giles de la Mare.
Kirby & Spence. 1843. *Introduction to Entomology; or, Elements of the Natural History of Insects: Comprising an Account of Noxious and Useful Insects*. London: Longman, Brown, Green, and Longmans.
Kirk, G.S. 1970. *Myth, Its Meaning and Functions in Ancient and Other Cultures*. Berkeley, CA: University of California Press.
Kizu, J., Neuman, C., Grand, L.L., and Liu, W. 2019. "Discovery of Cocirculating Ross River Virus and Barmah Forest Virus at Wide Bay Military Training Area, Northeastern Australia." *Journal of the American Mosquito Control Association* 35: 220–3.
Klausnitzer, B. 1981. *Beetles*. New York: Exeter Books.
Klein, B.A. 2012. "The Curious Connection Between Insects and Dreams." *Insects* 3: 1–17.
Klonsky, M. 1977. *William Blake: The Seer and His Visions*. New York: Harmony Books.
Konishi, M. and Ito, Y. 1973. "Early Entomology in East Asia." In R.F. Smith, T.E. Mittler, and C.N. Smith (eds.), *History of Entomology*, 1–20; facing p. 4. Palo Alto, CA: Annual Reviews, Inc.
Kors, A.C. 2002. *Encyclopedia of the Enlightenment*. Oxford: Oxford University Press.
Kritsky, G. 2010. *Quest for the Perfect Hive*. New York: Oxford University Press.
Kritsky, G. 2015. *Tears of Re; Beekeeping in Ancient Egypt*. New York: Oxford University Press.
Kritsky, G. 2018. "The Quest for The Perfect Hive: Ancient Mediterranean Origins." In F. Hatjina, G. Mavrofridis, and R. Jones (eds.), *Beekeeping in the Mediterranean from Antiquity to the Present*, 50–5. Nea Moudania: Division of Apiculture, Hellenic Agricultural Organization.
Kritsky, G. and Cherry, R. 2000. *Insect Mythology*. San Jose, CA: Writers Club Press.
Kroner, L. 1999. *Linnaeus: Nature and Nation*. Cambridge, MA: Harvard University Press.
Kuhn, A.J. 1956. "English Deism and the Development of Romantic Mythological Syncretism." *PMLA* 71(5): 1094.
Kuhn, K.G., Campbell-Lendrum, D.H., Armstrong, B., and Davies, C.R. 2003. "Malaria in Britain: Past, Present, and Future." *PNAS* 100: 9997–10001.
Kutschera, U. 2017. "Maria Sibylla Merian (1647–1717)." *Biologie in unserer Zeit* 47(1): 28–36.
Lai, D.H., Hashimi, H., Lun, Z.R., Ayala, F.J., and Lukes, J. 2008. "Adaptations of *Trypanosoma brucei* to Gradual Loss of Kinetoplast DNA: *Trypanosoma equiperdum* and *Trypanosoma evansi* are Petite Mutants of *T. brucei*." *PNAS* 105: 1999–2004.
Lamb, C. and Lamb, M. 1910. *Tales from Shakespeare*. New York: E. P. Dutton & Co.
Landin, B.-O. 1971. "Johann Christian Fabricius (1745–1808), Entomology." *Dictionary of Scientific Biography* 4: 512–13.

Landin, B.-O. 1972. "Charles de Geer (1720–78), Entomology." *Dictionary of Scientific Biography* 5: 328–9.

Lane, N. 2015. "The Unseen World: Reflections on Leeuwenhoek (1677) 'Concerning little animals'." *Philosophical Transactions of the Royal Society B* 370(1666). Available online: https://doi.org/10.1098/rstb.2014.0344 (accessed May 17, 2023).

Latham, M.W. 1930. *The Elizabethan Fairies: The Fairies of Folklore and the Fairies of Shakespeare*. New York: Columbia University Press.

Laursen, J.C., Nederman, C.J., and Hunter, I. 2005. *Heresy in Transition: Transforming Ideas of Heresy in Medieval and Early Modern Europe (Catholic Christendom, 1300–1700)*. Farnham: Ashgate Publishing Group.

Leeuwenhoek, A. 1700. "A Specimen of Some Observations Made with a Microscope." In *Royal Society Philosophical Transactions* 8: 6037–8, 6116–19 + plates 1939–9. Collected Letters. 17 vols, vol. 1, 29–39. Amsterdam: Swets & Zeitlinger.

Leith, J.A. 1971. "Peter Gay's Enlightenment." *Eighteenth-Century Studies* 5(1): 157.

Lescot, M., Audic, S., Robert, C., Nguyen, T.T., Blanc, G., Cutler, S.J., Wincker, P., Couloux, A., Claverie, J.M., Raoult, D., and Drancourt, M. 2008. "The Genome of *Borrelia recurrentis*, the Agent of Deadly Louse-Borne Relapsing Fever, is a Degraded Subset of Tick-Borne *Borrelia duttonii*." *PLoS Genetics*. Available online: https://doi.org/10.1371/journal.pgen.1000185 (accessed May 17, 2023).

Lindeboom, G.A., ed. and trans. 1975. *The Letters of Jan Swammerdam to Melchisédech Thévenot*. Amsterdam: Swets & Zeitlinger.

Lindroth, C.H. 1973. "Systematics Specializes Between Fabricius and Darwin." In R.F. Smith, T.E. Mittler, and C.N. Smith (eds.), *History of Entomology*, n. 2, 122–3. Palo Alto, CA: Annual Reviews, Inc.

Link, V.B. 2005. "A History of Plague in the United States of America." *Public Health Monograph* 26: 1–11.

Livio, M. 2020. "Did Galileo Truly Say, 'And Yet it Moves'? A Modern Detective Story." *Scientific American*. Available online: https://blogs.scientificamerican.com/observations/did-galileo-truly-say-and-yet-it-moves-a-modern-detective-story/ (accessed May 17, 2023).

Lomahaftewa, G. n.d. "Hopi Butterfly Dance – Circle of Dance." October 6, 2012 through October 8, 2017, The National Museum of the American Indian in New York. Available online: https://americanindian.si.edu/exhibitions/circleofdance/hopi.html (accessed May 17, 2023).

Lorange, E.A., Race, B.L., Sebbane, F., and Hinnebusch, J.B. 2005. "Poor Vector Competence of Fleas and the Evolution of Hypervirulence in *Yersinia pestis*." *Journal of Infectious Disease* 191: 1907–12.

Lyon, J. and Sloan, P.R. 1981. *From Natural History to the History of Nature: Readings from Buffon and His Critics*. South Bend, IN: University of Notre Dame Press.

Mackenzie, J.S., Lindsay, M.D., Coelen, R.J., Broom, A.K., Hall, R.A., and Smith, D.W. 1994. "Abroviruses Causing Human Disease in the Australian Zoogeographic Region." *Archives of Virology* 136: 447–67.

Mägdefrau, K. 1978. "Rudolph Jakob Camerarius (Camerer, 1665–1721), Medicine, Botany." *Dictionary of Scientific Biography* 15: 67–8.

Majanlahti, A. 2006. *The Families Who Made Rome: A History and a Guide*. London: Pimlico.

Malpighi, M. and Adelmann, H.B. 1975. *The Correspondence of Marcello Malpighi*. Ithaca, NY: Cornell University Press.

Mandelbrote, S. 2004. "John Ray [formerly Wray] (1627–1705), Naturalist and Theologian." *Oxford Dictionary of National Biography* 46: 178–83.
Mandeville, B. 1924. *The fable of the bees: or, private vices, publick benefits*, 2 vols. Oxford: Clarendon Press.
Manley, F. 1961. "Milton and the Beasts of the Field." *Modern Language Notes* 76(5) 398–403.
Manson, P. 1883. *The filaria Sanguinis hominis and certain new parasitic diseases in India, China, and certain warm countries*. London: H.K. Lewis.
Marr, J.S. and Cathey, J.T. 2013. "The 1802 Saint-Domingue Yellow Fever Epidemic and the Louisiana Purchase." *Journal of Public Health Management and Practice* 19(1): 77–82.
Mathews, C. 2009. *Modern Satanism: Anatomy of a Radical Subculture*. Santa Barbara, CA: Greenwood Publishing Group.
Matsuya Piece-Goods Store. 1972. *Japanese Design Motifs*. New York: Dover.
Mauriès, P. 2002. *Cabinets of Curiosities*. London: Thames & Hudson.
Mayr, E. 1982. *The Growth of Biological Thought: Diversity, Evolution, and Inheritance*. Cambridge, MA: Belknap Press.
McDade, J.E. 1987. "Flying Squirrels and their Ectoparasites: Disseminators of Epidemic Typhus." *Parasitology Today* 3(3): 85–7.
Meli, D.B. 1997. *Marcello Malpighi: Anatomist and Physician*. Florence: Leo S. Olsckhi.
Meli, D.B. 2000. "Marcello Malpighi (1628–94)." In W. Applebaum (ed.), *Encyclopedia of the Scientific Revolution from Copernicus to Newton*, 388–90. New York: Garland.
Meli, D.B. 2011. *Mechanism, Experiment, Disease: Marcello Malpighi and Seventeenth-Century Anatomy*. Baltimore, MD: Johns Hopkins University Press.
Melnikoff, N. 1869. "Uber die Jugendzustande der *Taenia cucumerina*." *Archiv für Naturgeschichte* 35: 62–70.
Merian, M.S. and O'Brien-Twohig, S. 2012. "From Metamorphosis insectorum Surinamensium (Or, Transformations of Surinamese Insects)." *Review: Literature and Arts of the Americas* 45(1): 21–8.
Mikkola, K. 1985. "The Geometroidea and Noctuidea Described by Carl Clerck (Lepidoptera)." *Insect Systematics and Evolution* 16(2): 121–9. Available online: https://doi.org/10.1163/187631285X00207 (accessed May 17, 2023).
Miller, G.L. 1997. "Historical Natural History: Insects and the Civil War." *American Entomologist* 43: 227–45.
Miller, L.H., Mason, S.J., Clyde, D.F., and McGinniss, M.H. 1976. "The Resistance Factor to *Plasmodium vivax* in Blacks." *The New England Journal of Medicine* 295: 302–4.
Miller, W.S. 1935. "Elisha Perkins and His Metallic Tractors." *Yale Journal of Biology and Medicine* 8: 41–57.
Milton, J. [1667] 2005. *Paradise Lost*. New York/London: W.W. Norton & Company.
Mitford, N. 2012. *Voltaire in Love*. New York: NYRB Classics.
Montalenti, G. 1976. "Antonio Vallisneri (1661–1730), Biology, Medicine." *Dictionary of Scientific Biography* 13: 562–5.
Morand, S., McIntyre, K.M., and Baylis, M. 2014. "Domesticated Animals and Human Infectious Diseases of Zoonotic Origins: Domestication Time Matters." *Infection, Genetics and Evolution* 24: 76–81.
Muffet, T. 1634. *Insectorum sive minimorum animalium theatrum*. London.

Munz, T. 2016. *The Dancing Bees: Karl Von Frisch and the Discovery of the Honeybee Language*. Chicago, IL: University of Chicago Press.

Murtagh, A. and Lee, A.2017. *Meta: On God, the Big Questions, and the Just City (An Uncommon Exchange)*. Eugene, OR: Cascade Books.

Museum of Comparative Zoology, Harvard University. 2019. Available online: https://mcz.harvard.edu/ (accessed September 28, 2019).

Mushtaq, M.U. 2009. "Public Health in British India: A Brief Account of the History of Medical Services and Disease Prevention in Colonial India." *Indian Journal of Community Medicine* 34(1): 6–14.

Nauenberg, M. 1994. "Hooke, Orbital Motion, and Newton's Principia." *American Journal of Physics* 62: 331–50.

Nelson, L. n.d. "The Great Famine (1315–1317) and the Black Death (1346–1351)." Available online: http://www.vlib.us/medieval/lectures/black_death.html (accessed May 17, 2023).

Neri, J. 2011. *The Insect and the Image: Visualizing Nature in Early Modern Europe 1500–1700*. Minneapolis, MI: University of Minnesota Press.

Panofsky, E. 1966. *Early Netherlandish Painting: Its Origins and Character*, vol. 1. Cambridge, MA: Harvard University Press.

Porny, M.A. 1795. *The Elements of Heraldry, Containing the Definition, Origin, and Historical Account of That Ancient, Useful, and Entertaining Science*, 5th edn. London.

Odorico, D.M., Graves, S.R., Currie, B., Catmull, J., Nack, Z., Ellis, S., Wang, L., and Miller, D.J. 1998. "New *Orientia tsutsugamushi* Strain from Scrub Typhus in Australia." *Emerging Infectious Diseases* 4(4): 641–3.

Ogilvie, B.W. 2012. "Attending to Insects: Francis Willughby and John Ray." *Notes & Records of the Royal Society* 66(4): 357–72.

Olaquiaga, C. 2005–2006. "Object Lesson/Transitional Object, Cabinet, Winter 2005–06." Available online: http://www.cabinetmagazine.org/issues/20/olalquiaga.php (accessed May 17, 2023).

Olbricht, E. 2006. "Made Without Hands: The Representation of Labor in Early Modern Silkworm and Beekeeping Manuals." In E.C. Brown (ed.), *Insect Poetics*, 223–41. Minneapolis/London: University of Minnesota Press.

Olby, R. 1966. *Origins of Mendelism*. New York: Schocken Books.

Olby, R. 1973. "Joseph Gottlieb Koelreuter (1733–1806), Botany." *Dictionary of Scientific Biography* 7: 440–2.

Ordish, G. 1976. *The Constant Pest: A Short History of Pests and their Control*. New York: Charles Scribner's Sons.

Orihel, T.C. and Lowrie, Jr., R.C. 1975. "*Loa loa*: Development to the Infective Stage in an American Deerfly, *Chrysops atlanticus*." *The American Journal of Tropical Medicine and Hygiene* 24(4): 610–15.

Ørsted, H.C. 1799. "Fundamentals of the Metaphysics of Nature Partly According to a New Plan." *Philosophisk Repertorium*. Printed by Boas Brünnich, Copenhagen, in Danish. Kirstine Meyer's 1920 edition of Ørsted's works, vol. I, pp. 33–78. English translation by K. Jelved, A.D. Jackson, and O. Knudsen (1997).

Outram, D. 2019. *The Enlightenment*. Cambridge: Cambridge University Press.

Outram, Q. 2001. "The Socio-Economic Relations of Warfare and the Military Mortality Crises of the Thirty Years' War." *Medical History* 45: 151–84.

Owen, J. 2017. *The Feminine Monarchie*. Yorkshire, UK: Northern Bee Books.

Paget, S. 1897. *John Hunter, Man of Science and Surgeon (1728–1793)*. London: T. Fischer Unwin.

Palm, L.C. 2000. "Antoni van Leeuwenhoek (1632–1723)." In W. Applebaum (ed.), *Encyclopedia of the Scientific Revolution from Copernicus to Newton*, 609–14. New York: Garland.

Panzer, G.W. [1796] 1813. Faunae Insectorum Germanicae Initia, oder Deutschlands Insecten, 109 Teile, 2640 Kupfertafeln von Jacob Sturm, herausgegeben von Dr G. W. F. Panzer. Zweyte Auflage. [Fortgesetzt bis 1844 von] Dr G. A. W. Herrich-Schaffer.

Parker, P. 2001. *Europe in Crisis: 1598–1648*. Hobeken, NJ: Wiley-Blackwell.

Paterson, W.J. 1956. "Benjamin Franklin, the Freemason." *Journal of the Franklin Institute* 261(1): 107–9.

Patterson, K.D. 1992. "Yellow Fever Epidemics and Mortality in the United States, 1693–1905." *Social Science and Medicine* 34(8): 855–65.

Patterson, K.D. 1993. "Typhus and its Control in Russia, 1870–1940." *Medical History* 37: 361–81.

Patterson, R. 1838. *Letters on the Natural History of the Insects Mentioned in Shakespeare's Plays, with Incidental Notices of the Entomology of Ireland*. London: Wm. S. Orr and Co., Paternoster Row.

Pauly, P.J. 2002. "Fighting the Hessian Fly: American and British Responses to Insect Invasion; 1776–1789." *Environmental History* 7(3): 485–507.

Pearson, R. 2005. *Voltaire Almighty: A Life in Pursuit of Freedom*. London: Bloomsbury.

Peschke, Z. 2007. "The Impact of the Black Death." *ESSAI* 5(32).

Pesic, P. 1999. "Wrestling With Proteus: Francis Bacon and the "torture" of Nature." *Isis* 1: 81–94.

Peterson, R.K.D. 1995. "Insects, Disease, and Military History: The Napoleonic Campaigns and Historical Perception." *American Entomologist* 41: 147–60.

Pilet, P.E. 1970. "Charles Bonnet (1720–93), Natural History, Biology, Philosophy." *Dictionary of Scientific Biography* 2: 286–7.

Powell, J.R. and Tabachnick, W.J. 2013. "History of Domestication and Spread of *Aedes aegypti* – A Review." *Memórias do Instituto Oswaldo Cruz* 108(Supplement 1): 11–17.

Pruett, J. 1963. "Charles Butler—Musician, Grammarian, Apiarist." *The Musical Quarterly* 49(4): 498–509.

Raoult D. and Roux, V. 1999. "The Body Louse as a Vector of Reemerging Human Diseases." *Clinical Infectious Disease* 29(4): 888–911.

Raoult, D., Aboudharam, G., Crubezy, E., Larrouy, G., Ludes, B., and Drancourt, M. 2000. "Molecular Identification by 'Suicide PCR' of *Yersinia pestis* as the Agent of the Medieval Black Death." *PNAS* 97: 12800–3.

Raoult, D., Dutour, O., Houhamdi, L., Jankauskas, R., Fournier, P., Ardagna, Y., Drancourt, M., Signoli, M., La, V.D., Macia, Y., and Aboudharam, G. 2006. "Evidence for Louse-Transmitted Diseases in Soldiers of Napoleon's Grand Army in Vilnius." *Journal of Infectious Diseases* 193(1): 112–20.

Raoult, D., Reed, D.L., Dittmar, K., Kirchman, J.J., Rolain, J., Guillen, S., and Light, J.E. 2008. "Molecular Identification of Lice From Pre-Columbian Mummies." *Journal of Infectious Diseases* 197: 535–43.

Rappaport, R. 1991. "Italy and Europe: The Case of Antonio Vallisneri (1661–1730)." *History of Science* 29: 73–98.

Rasmussen, S.A., Jamieson, D.J., Honein, M.A., and Petersen, L.R. 2016. "Zika Virus and Birth Defects – Reviewing the Evidence for Causality." *New England Journal of Medicine* 374: 1981–7.

Ratcliffe, M.J. 2009. "The Quest for the Invisible: Microscopy in the Enlightenment." New York: Routledge.

Raven, C.E. 1942. *John Ray, Naturalist: His Life and Works*. Cambridge: Cambridge University Press.

Raven, P.H., Berlin, B., and Breedlove, D.E. 1971. "The Origins of Taxonomy." *Science* 174 (4015): 1210–13.

Ray, J. 1660. *Catalogus plantarum circa Camtabrigiam nascentium*. Cambridge: Jeann. Field.

Ray, J. 1710. *Historia Insectorum*. London: A. & J. Churchill.

Ray, J. 1975. *Flora of Cambridgeshire*. Trans. A.H. Ewen and C.T. Prime. Hitchin: Wheldon & Wesley.

Réaumur. 1926. *The Natural History of Ants*. Ed. and trans. W.M. Wheeler. New York: Knopf. New York: Arno Press, 1977.

Réaumur. 1955. *Histoire des scarabées*. Introduction by M. Caullery. Paris: Paul Lechevalier

Redi. 1668. *Esperienze intorno alla generazione degli insetti, scritte in una lettera a Carlo Dati*. Florence.

Redi. 1909. *Experiments on the Generation of Insects*. Trans. M. Bigelow. Chicago: Open Court Publishing Co.

Reel, M. 2013. *Between Man and Beast: An Unlikely Explorer and the African Adventure that took the Victorian World by Storm*. New York: Anchor.

Rees, N. 2003. "Note that Jefferson is identified as 'American Polymath and President'." In *Cassell's Humorous Quotations*, 392. New York: Sterling Publishing Company.

Reeves, W.K., Utter, C.M., and Durden, L.A. 2012. "Rickettsial Pathogens and Arthropod Vectors of Medical and Veterinary Significance on Kwajalein Atoll and Wake Island." *Micronesica* 43: 107–13.

Reilly, P.C. 1974. *Athanasius Kircher S. J.: Master of a Hundred Arts*. Wiesbaden: Edizioni del Mondo.

Reiter, P. 2000. "From Shakespeare to Defoe: Malaria in England in the Little Ice Age." *Emerging infectious Diseases* 6: 1–11.

Repcheck, J. 2008. *Copernicus' Secret: How the Scientific Revolution Began*. New York: Simon & Schuster.

Richard, G. 1973. "The Historical Development of Nineteenth and Twentieth Century Studies on the Behavior of Insects." In R.F. Smith, T.E. Mittler, and C.N. Smith (eds.), *History of Entomology*, n. 2, 477–502. Palo Alto, CA: Annual Reviews, Inc.

Richardson, A. 1802. Abijah Richardson to Benjamin Waterhouse, 1 February 1802, Box 1, Folder 2, Benjamin Waterhouse papers, 1782–1841. David M. Rubenstein Rare Book & Manuscript Library, Duke University, Durham, NC.

Robinson, D. 2007. "The Bugs that Bugged the Colonists." Colonial Williamsburg, Autumn.

Roscoe, W. 1808. *The Butterfly's Ball and the Grasshopper Feast*. London.

Rousseau, J.-J. 1755. *Discours sur l'origine et les fondements de l'inégalité parmi les hommes* (Discourse on the Origin and Basis of Inequality Among Men). Holland: Marc-Michel Rey.

Routh, H.B. and Bhowmik, K.R. 2007. "History of Elephantiasis." *International Journal of Dermatology* 32(12): 913–16.

Salmon, M.A. 2000. *The Aurelian Legacy: British Butterflies and Their Collectors*. London: Harley Books.

Sarton, G. 1943. "The Feminine Monarchie of Charles Butler." *Isis* 34(6).
Savioz, R. 1948. *Mémoires Autobiographiques de Charles Bonnet de Geneve*. Paris: J. Vrin.
Scally, A. and Durbin, R.2012. "Revising the Human Mutation Rate: Implications for Understanding Human Evolution." *Nature Reviews, Genetics* 13: 745–53.
Schachat, S.R. 2014. "Drawn Before Wonderland: Bizarre Illustrated Insects of the Nineteenth Century." *American Entomologist* 60(3): 162–5.
Schanzer, E. 1955. "The moon and the Fairies in *A Midsummer Night's Dream*." *University of Toronto Quarterly* 24(3): 234–46.
Schierbeek, A. 1967. *Jam Swammerdam (12 February 1637–17 February 1680): His Life and Works*. Amsterdam: Swets & Zeitlinger.
Schmidt, L.H., Fradkin, R., Harrison, J., and Rossan, R.N. 1977. "Differences in the Virulence of *Plasmodium knowlesi* for *Macaca irus* (*fascicularis*) of Philippine and Malayan Origins." *The American Journal of Tropical Medicine and Hygiene* 26(4): 612–22.
Schrader, D.V. 1962. "The Newton-Leibniz Controversy Concerning the Discovery of the Calculus." *The Mathematics Teacher* 55(5): 385–96.
Scot, R. 1584. *The Discoverie of Witchcraft*. London.
Scott, F.M. 1927. "The Botany of Marcello Malpighi, Doctor of Medicine." *Scientific Monthly* 25(6): 546–53.
Scribner, R.W. 1993. "The Reformation, Popular Magic, and the 'Disenchantment of the World'." *The Journal of Interdisciplinary History* 23(3).
Shachtman, T. 2014. *Gentlemen Scientists and Revolutionaries: The Founding Fathers in the Age of Enlightenment*. New York: St. Martin's Press.
Shakespeare, W. 1908. *A Midsummer-Night's Dream*. New York: Doubleday.
Shakespeare, W. [1599–1600] 2016. *Henry V*. In S. Greenblatt et al. (ed.), *The Norton Shakespeare*. New York/London: W.W. Norton & Company.
Shakespeare, W. [1600] 2016. *Hamlet*. In S. Greenblatt et al. (ed.), *The Norton Shakespeare*. New York/London: W.W. Norton & Company.
Shakespeare, W. [1601] 2016. *Twelfth Night*. In S. Greenblatt et al. (ed.), *The Norton Shakespeare*. New York/London: W.W. Norton & Company.
Shakespeare, W. [1605] 2016. *King Lear*. In S. Greenblatt et al. (ed.), *The Norton Shakespeare*. New York/London: W.W. Norton & Company.
Shakespeare, W. [1606] 2016. *Macbeth*. In S. Greenblatt et al. (ed.), *The Norton Shakespeare*. New York/London: W.W. Norton & Company.
Shakespeare, W. [1606–1607] 2016. *Pericles*. In S. Greenblatt et al. (ed.), *The Norton Shakespeare*. New York/London: W.W. Norton & Company.
Shakespeare, W. [1607–1608] 2016. *Antony and Cleopatra*. In S. Greenblatt et al. (ed.), *The Norton Shakespeare*. New York/London: W.W. Norton & Company.
Shakespeare, W. [1609] 2016. *Cymbeline*. In S. Greenblatt et al. (ed.), *The Norton Shakespeare*. New York/London: W.W. Norton & Company.
Shakespeare, W. [1609] 2016. *The Winter's Tale*. In S. Greenblatt et al. (ed.), *The Norton Shakespeare*. New York/London: W.W. Norton & Company.
Sharp, C.C. 2001. "Nature's Laws Revealed in Rhyming Couplets." *Nature* 413. Available online: https://doi.org/10.1038/35093273 (accessed May 17, 2023).
Sharratt, M. 1994. *Galileo: Decisive Innovator*. Cambridge: Cambridge University Press.
Shelley, P.B. [1819] 2010. "Song: To the Men of England." In S.C. Behrendt (ed.), *A Longman Cultural Edition: Percy Bysshe Shelley*, 19–20. London: Pearson.

Shelley, P.B. 1820. *Oedipus Tyrannus or Swellfoot the Tyrant, a Tragedy*. London: J. Johnston.

Shelley, P.B. [1813] 1831. *Queen Mab*. New York: Wright & Owen.

Shuker, K. 2017. "The Locust of Kalisz? More Like a Dali-esque Deathshead!" *Shukernature*. Available online: http://karlshuker.blogspot.com/2017/07/the-locust-of-kalisz-more-like-dali.html (accessed May 17, 2023).

Siebmacher, J.A. 1906. *Grosses und allgemeines wappenbuch, Bd. 3, Abth. 2, B. 2*, vol. 3, ch. 2, pt. 2. Nürnberg, Germany: Bauer and Raspe.

Sill, G.G. 1975. *A Handbook of Symbols in Christian Art*. New York: Macmillan.

Simpkins, D.M. 1974. "Thomas Moffett (Moufet, Muffet, 1553–1604), Medicine, Entomology." *Dictionary of Scientific Biography* 9: 440–1.

Singer, C. 1953. "The Earliest Figures of Microscopical Objects." *Endeavour* 12: 197–201.

Sleigh, C. 2006. "Inside Out: The Unsettling Nature of Insects." In E.C. Brown (ed.), *Insect Poetics*, 281–97. Minneapolis/London: University of Minnesota Press.

Sloan, A.W. 1973. "Medical and Social Aspects of the Great Plague of London in 1665." *South African Medical Journal* 47: 270–6.

Sluhovsky, M. 2007. *Believe Not Every Spirit: Possession, Mysticism, & Discernment in Early Modern Catholicism*. Chicago, IL: The University of Chicago Press.

Smith, C.U.M. and Arnott, R. 2005. *The Genius of Erasmus Darwin*. Aldershot: Ashgate.

Smith, D.C. 1980. "Gerhard's Distinction Between Typhoid And Typhus and its Reception in America, 1833–1860." *Bulletin of the History of Medicine* 54(3): 368–85.

Sneli, J.-A., Knudsen, J., and Vedelsby, A.2009. "Johan Christian Fabricius and his Molluscan Species, *Acesta excavata* (J. C. Fabricius, 1779)." *Steenstrupia* 30(2): 153–62.

Snorrason, E. 1974. "Otto Frederik Müller (1730–84), Botany, Zoology." *Dictionary of Scientific Biography* 9: 574–6.

Sonar, T. 2018. *The History of the Priority Dispute between Newton and Leibniz: Mathematics in History and Culture*. Heidelberg: Springer Nature.

Soulsby, B.H. 1933. *A catalogue of the works of Linnæus (and publications more immediately relating thereto) preserved in the libraries of the British Museum (Bloomsbury) and the British Museum (Natural History)*. London: Printed by order of the Trustees.

Southerne, E. 1593. *A Treatise concerning the right use and ordering of bees: newlie made and set forth, according to the author's owne experience: (which by any heretofore hath not been done)*. Thomas Owen for Thomas Woodcocke, London.

Southwood, T.R.E. 1977. "Entomology and Mankind: Insects Over the Ages Have Greatly Affected Man's Health and Food Supply and Have Played an Important Role as Religious and Cultural Symbols." *American Scientist* 65(1).

Spärck, R. 1932. "Otto Friedrich Müller, 1730–1784." In V. Meisen (ed.), *Prominent Danish Scientists Through the Ages*, 60–4. Copenhagen: Levin & Munksgaard.

Spary, E.C. 2004. "Scientific Symmetries." *History of Science* 42: 9.

Spenser, E. [1590] 2014. *A Letter of the Authors. Edmund Spenser's Poetry*. New York/London: W.W. Norton & Company.

Squires, V.P. 1894. "Milton's Treatment of Nature." *Modern Language Notes* 9(8): 227–37.

Staloff, D. 2007. *Hamilton, Adams, Jefferson: The Politics of Enlightenment and the American Founding*. New York: Hill and Wang.

Starkey, A. 2003. *War in the Age of Enlightenment, 1700–1789 Studies in Military History and International Affairs*. Santa Barbara, CA: Greenwood Publishing Group.

Steffens, H. 1842. *Was Ich Erlebte: Aus Der Erinnerung Niedergeschrieben Von Henrich Steffens*. Facsimile Publishing.

Steverding, D. 2008. "The History of African Trypanosomiasis." *Parasites and Vectors* 1: 3.

Steverding, D. 2017. "The History of Leishmaniasis." *Parasites and Vectors* 10: 82.

Swift, J. 2010. *The Battle of the Books and Other Works, Including "A Modest Proposal."* Floating Press, from an 1886 edition.

Thalassinou, E., Tsiamis, C., Poulakou-Rebelakou, E., and Hatzakis, A.2015. "Biological Warfare Plan in the 17th Century—the Siege of Candia, 1648–1669." *Emerging Infectious Disease* 21(12): 2148–53.

The Linnean Society of London. 2019. Available online: https://www.linnean.org/the-society/history-of-science (accessed September 27, 2019).

The Royal Society. 2019. Available online: https://royalsociety.org/about-us/history/ (accessed September 27, 2019).

Thomas, K. 1971. *Religion and the Decline of Magic*. London: Weidenfeld and Nicolson.

Thorley, J. 1744. *Melisselogia or, the female monarchy*. London: N. Thorley.

Thornton, R. 1990. *American Indian Holocaust and Survival: A Population History since 1492*. Norman, OK: University of Oklahoma Press.

Todd, K. 2007. *Chrysalis: Maria Sibylla Merian and the Secrets of Metamorphosis*. Orlando, FL: Houghton Mifflin Harcourt.

Torlais, J. 1961. *Un Esprit Encyclopédique en dehors de "L'Encyclopédie" Réaumur d'après les documents inédits*, 2nd edn. Paris: Blanchard.

Trevisanato, S.I. 2007. "The Biblical Plague of the Philistines Now Has a Name, Tularemia." *Medical Hypotheses* 69: 1144–6.

Turnbull, H.W. 1960. *The Correspondence of Isaac Newton*, vol. II. Cambridge: Cambridge University Press.

Tuxen, S.L. 1967. "The Entomologist J.C. Fabricius." *Annual Review of Entomology* 12: 1–14.

Tuxen, S.L. 1973. "Entomology Systematizes and Describes: 1700–1815."In R.F. Smith, T.E. Mittler, and C.N. Smith (eds.), *History of Entomology*, n. 2, 95–118. Palo Alto, CA: Annual Reviews, Inc.

Valiant, S. 1993. "Maria Sibylla Merian: Recovering an Eighteenth-Century Legend." *Eighteen-Century Studies* 26(3): 467–79.

Vallisneri, A. 1713. *Esperienze, ed Obsservazioni intorno all' origine, sviluppi, e costume di varj insetti*. Padua, Gio: Manfrè.

Vallisneri, A. 1932. "Observations on the Fly of Rosebushes and Other Insects of the Same Habitat." Trans. M. Bigelow and R.B. Bigelow. *Isis* 17: 290–324.

Van de Roemer, B. 2004. "Neat Nature: The Relation Between Nature and Art in Dutch Cabinet of Curiosities from the Early Eighteenth Century." *History of Science* 42: 47–84.

Van de Roemer, G.M. 2014. "Redressing the Balance: Levinus Vincent's Wonder Theatre of Nature." *The Public Domain Review*. Available online: https://publicdomainreview.org/2014/08/20/redressing-the-balance-levinus-vincents-wonder-theatre-of-nature/ (accessed May 17, 2023).

Van de Wetering, E. 2005. "Rembrandt Leaning on a Windowsill." In Rembrandt Research Project Foundation, *A Corpus of Rembrandt Paintings*, vol. 4. Dordrecht: Springer.

van Leewenhoeck, A. 1677. "Observations, communicated to the publisher by Mr. Antony van Leewenhoeck, in a Dutch letter of the 9th of Octob. 1676. Here English'd: concerning little animals by him observed in rain-well-sea and snow water; as also in water wherein pepper had lain infused." *Philosophical Transactions* 12: 821–31.

Vignau-Wilberg, T. 2017. *Joris and Jacob Hoefnagel: Art and Science Around 1600*. Berlin, Germany: Hatje Cantz Verlag GmbH.

Vigué, J. 2002. *Great Women Masters of Art*. New York: Watson-Guptill Publications.

Von Engeln, O.D. 1916. "Shakespeare, the Observer of Nature." *The Scientific Monthly* 2(6): 573–88.

Voss, J.D., Kugblenu, R., Salter, K., Johnson, L., and Reeves, W.K. 2016. "Case Series of 23 Deaths from Hymenoptera Stings Among United States Air Force Populations." *Journal of Hymenoptera Research* 48: 95–9.

Wansink, B., Mukund, A., and Weislogel, A. 2016. "Food Art Does Not Reflect Reality: A Quantitative Content Analysis of Meals in Popular Paintings." *SAGE Open* 6(3): 1–10.

Weiss, H. 1926. "Two Entomologists of the Eighteenth Century—Eleazar Albin and Moses Harris." *The Scientific Monthly* 23(6): 558–64.

Weiss, H. 1930. "Insects and Witchcraft." *Journal of the New York Entomological Society* 38(2): 127–33.

Weiss, H.B. and Ziegler, G.M. 1931. *Thomas Say, Early American Naturalist*. Springfield, IL: C.C. Thomas.

Wellington, J. 1994. *Secondary Science: Contemporary Issues and Practical Approaches*. New York: Routlege.

Wellisch, H.H. 1984. *Conrad Gessner: A Bio-Bibliography*. Zug, Switzerland: IDC.

Weny, G., Okwee-Acai, J., Okech, S.G., Tumwine, G., Ndyanabo, S., Abigaba, S., and Goldberg, T.L. 2017. "Prevalence and Rick Factors Associated With Hemoparasites in Cattle and Goats at the Edge of Kibale National Park, Western Uganda." *Journal of Parasitology* 103: 69–74.

Werness, H.B. 2004. *The Continuum Encyclopedia of Animal Symbolism in Art*. New York: Continuum International Publishing.

Wheler, G. 1682. *A Journey into Greece*. London: W. Cademan and others.

White, S. 1759. *Collateral Bee-Boxes: Or, a New, Easy, and Advantageous Method of Managing Bees*. London: L. Davis and C. Reymers.

Wildman, T. 1770. *A Treatise on the Management of Bees*. London: W. Strahan and T. Cadell.

Wilkie, J.S. 1956a. "The Idea of Evolution in the Writings of Buffon.—II." *Annals of Science* 12(3): 212–27.

Wilkie, J.S. 1956b. "The Idea of Evolution in the Writings of Buffon.—III." *Annals of Science* 12(4): 255–66.

Wilkins, J.S. 2009. *Defining Species: A Sourcebook from Antiquity to Today*. New York: Peter Lang Inc., International Academic Publishers.

Williams, C.A.S. 1976. *Outlines of Chinese Symbolism; An Alphabetical Compendium of Antique Legends and Beliefs, as Reflected in the Manners and Customs of the Chinese*. New York: Dover Publication Inc.

Williams, H.S. 2018. *A History of Science*, vol. 4. Sagwan Press.

Wilson, D.L. 2004. "Edward Donovan (1768–1837)." *Dictionary of Nineteenth-Century British Scientists* 2: 604–5.

Winsor, M.P. 1976. "Jan Swammerdam (1637–1680), Biology." *Dictionary of Scientific Biography* 13: 168–75.
Woodruff, B.L. 1953. "Keats's Wailful Choir of Small Gnats." *Modern Language Notes* 68(4): 217–20.
Worthington, D. 2016. "Robert Hooke and the Wrath of Isaac Newton." Available online: https://www.newhistorian.com/2016/08/22/robert-hooke-wrath-isaac-newton/ (accessed May 17, 2023).
Wulf, A. 2015. "The Forgotten Father of Environmentalism." *The Atlantic*.
Zipes, J. 2002. *The Brothers Grimm: From Enchanted Forests to the Modern World*, 2nd edn. New York: Palgrave Macmillan.

INDEX